Curaçao, 2021

This book is written / compiled by
John H Baselmans-Oracle

Layout and front by
John H Baselmans-Oracle

With special thanks to those people who have helped me.

ISBN 978-1-008-91384-4

Time

John H Baselmans - Oracle

Index

Chapter 2   34

Science first

Chapter 7 137

Final

"If you want to find the secrets of the Universe,
Think in terms of energy, frequency and vibrations."

Nikola Tesla

Foreword

Two input frequencies, one low, one high
and the higher frequency must be 11 times the lower.
It's called "the 11th harmonic".

Before I start writing anything and explaining anything to you, I started again with that same quote from Nicola Tesla, as in my previous book. The most important statement for humanity and the answer to all healings and also what is going on in the current corona era! Life is frequencies and you could read about that in my previous book called "Quantiversum". We also know that frequencies can heal people as well as make them ill and even eliminate them! For a long time we have known that frequencies can cure people of any illness. One such man was Nikola Tesla, who faced scandal and death. Many more followed and now, 150 years later, it is impossible to escape the fact that many of man's secrets are in the world of frequencies.

We now know that we can heal people and make cancers disappear, but we also know that many of the frequencies are being used to regulate humanity, which is what we call 5G. Something that in the past was done sporadically and locally has now become a network where with the push of a button a person can be killed. This is already happening to several people who have been vaccinated with the experimental Nano technology.
Then wasn't that Nano technology so experimental?

In this book, we are all going to read it and we are really going to the core of life. Of course you are going to feel uncomfortable, perhaps angry and for sure helpless, but this is my way of telling you the truth that I receive from higher powers, and I will do so until my last day.

So let us start with this book.

TIME

Introduction

I have written a lot about it and quoted things that shocked many people and then, to cap it all, I will close this topic. This book will be about the greatest misconception that has arisen in mankind.

The word "time" is what determines, regulates and keeps mankind under its spell!
It makes science get nowhere and is full of limitations.
It keeps us stuck in a world of numbers that are straying from reality.
But it also makes that statements are seen as true, just because they are based on the given time and thus distance.

Let's unravel time and dissect it in its entirety and then we will come to a very painful conclusion:

"There is no time, there is no distance"!

Chapter 1

Some notes from my book "World of positive energy"

1-1
First

First I want to to come up with this piece because it explains what today's humanity is and how they hold on to a system that is NOT there for the people but for power.

Why do we try so hard to turn everything upside down?
Why don't we learn from this pure life form, which is here to finish its job? It is incredible what the world does. We always think that we have to do this or that, but what we really do is what other people tell us to do. All the time, we look at the world around to be sure we do the right thing. Why? What is it that makes people so afraid that they do what other people want them to do? Over the centuries, human life has been dominated by a system made by powerful people who pretend to know everything. This system has its rules, rules made by those who want to have power over others. It is clear all over the world, including the US and Europe. These systems are no longer human, they are controlled by computers and digits. If you do not have the right documents, you are outside the system. So every second of our life goes into that system. We cannot think or listen to our feelings; there is no time. The system kills our feelings, making us more dead than alive. What does such a human being do for the Universe in his lifetime? NOTHING! Such human beings produce nothing. This is the reason why they come back time and again, back to this earth to do their tasks all over again. Humans today are like sheep; they follow one dominant character and do not look to the right or to the left. They are afraid to be thrown out of the system, to be put in jail, to be made brain-dead or simply eliminated. This is what happens if you think too far outside the system. Look around you. People who protest against the system are gone after a while. The

system determines what you can do. Every time I hear or read the phrase 'This is a free country', I can't help laughing. Free, who knows what freedom means? Nobody can do anything outside the system. There is no such thing as a free country! There is no freedom on earth. Because if you refuse to participate in the system, you are considered sick, mad or crazy.

1-2
Then I came to this statement (back to dust)

It really tells us what time is and that time is determined, not by a definite fact, but a determination made by a system.

For a hill to become dust takes millions of years. But in a human life, this cycle is about 80 years. The difference is just time. We are all born strong and healthy. Our Universal energy is 100%. As a teenager, we start going downhill, because our food is unhealthy, we refuse to listen to our body, and we do only what we like to do. As we get older, we feel that something has gone wrong. Our body tells us: "I stop, I quit". We feel as if we're torn apart, because our body no longer works like before. Eventually, the body gives up and turns to dust, becoming an integrated part of the earth.

1-3
Healing in time

You will see that time heals all problems and you will see what others were trying to do to you. Being emotional is no longer a problem and others cannot use it against you.

1-4
Travel in time

As we know now, the soul can be everywhere. It comes and goes. A soul can live in different bodies during one human life. As limitations of time and distance do not exist, the soul utilizes the time when we do not use it. For example, when we sleep or when we do our job (on the automatic pilot), it can go to another body that needs its input. Because of influences from outside, the soul decides to be there. Then it is temporarily away from our body. One soul can live in several human bodies. It is not a problem, but it depends on what kinds of tasks it has. All souls have their own tasks to fulfill. How they accomplish their tasks, depends on how strong they are. A young soul that has just started, has more problems coping with its first task. Therefore, it needs all of its energy. An older soul can be a part of a human life with only 20% of its energy or less and so uses the other 80% in other bodies. Normally, 50% of the main power stays in the soul world to manage those different lives.

1-5
Time and the soul or in other words your life

As you see, "ghosts" are not lost. They search for a new body and sometimes it is difficult for them to find the right one. As I told you, time is not a problem and a ghost that is hundreds years old, for example, is a "newborn" in a soul's life. Every soul returns to the soul's world. In the Universe, there is a "place" where all souls come together and wait for their next task. They are all connected to each other, because they work together for the same Universe.

1-7
Virtual life or the in other words, the current system

I can almost hear you thinking and I can guess the questions you have right now. This is no science fiction, but exactly how a human life works. With all our technology and our knowledge, we are not capable of finding sources. Nor can we get answers for the questions about the source of life, or about the origins of the earth. These parts seem as if they do not exist, or we humans cannot find the key. When we think we have found it, we come upon other worlds behind the last one. In the world of souls, there is no beginning and no end. There is no time, no dimension, because those only exist in the virtual life we live in. Why do I call this life a virtual one? Because the entire system is based on lies about a life that is not real. Indeed, the way we live today is like a movie, a computer program, or a soap opera. We even change our bodies and manipulate them. Everything is based on "nothing", without a real foundation. A soul can do everything it wants, but all souls must fulfill a task. In that way, they reach a higher level than we, humans, can imagine. There is no need to know either, because we cannot handle that information.

1-8
Travel with your body

The main source or part of your body is your soul. A soul has many lives, unthinkable in earthly timelines. A soul can be millions of years old and still learning and creating. The human body is only a small tool, which will be replaced repeatedly, so the soul can go on learning. The human body will be gone after the task is fulfilled. This happens when the body has reached the stage when it is not capable of going on. The soul leaves the body and continues with a new task in another one. All the knowledge you have gained and experienced in that body (life) remains in that soul forever. It will never be erased from that soul's "hard disk". The human body returns to dust and becomes part of this earth again, waiting for the day that it is a part of the sand, rock or food for

1-9
Not one but more

We are here many times, in several bodies. A human life is like a holiday! A human life is to be here and to make the best of all situations you come across on your path.

1-10
Time is energy

It is true that we cannot control things and so they keep happening. See it as a learning process and see after an incident how you reacted. Start now and when there is something unpleasant, get over it as soon as possible. If it is nice, enjoy it! In the world of energy, there is no time, no beginning and no end. There is energy you should work with, and you can do it as long as or as short as you want because…. What is time? Everything you see is energy, you can try to see its pixels as light or as a matrix. Energy is a pure source and what we do is spending billions to find the source of life. We forget that we already have the answer inside us. Try to feel the energy, but more importantly, try to listen to the energy and use it. We are not able to manipulate this energy and we are just starting to explore it. Do not waste your time or wrack your brain trying to understand it. Use it and listen to the energy that has only one path and that is the path of health and understanding. You will see that you can make your life paradise.

1-11
Conclusion

For a soul, there is no time, no distance and no limitations. The first two: time and distance are known by the soul at the moment it starts to exist. The soul should learn and find out how to work with limitations. The older a soul is, the more it knows what is possible in the soul world and so it will continue to climb higher until it has fulfilled its task. There is a lot to learn, to find out how to work without limitations. So, the soul needs to go to these worlds, to find out how to manage limitations. There is a lot more going on in this life here on earth. This is why we are here with all the limitations of time, distance and things that WE THINK exist! It is a world created by ourselves and we make it difficult for ourselves. The world is getting more complicated every day. When I see simple things explained in books of hundred pages or more, I know we have taken a wrong turn.

1-12
Explanation

Let me elaborate. We humans try to explain everything in numbers, rules, time and distances. With these limitations, we research everything and find out that behind every cell, there is another one. Behind every world another world, simply because the limitations we live with, keep us in a world that will never end. There is no end, so why try to find it? Look at earth as a grain of dust that is traveling in a desert. At that moment, we may think that we are lost and there is no end anywhere on earth. It looks as if the world is endless. But when the grain of dust enters the Universe, there is no number that can place or explain that. So what we are doing here, is learning to accept. Accept that there is no end and that we, as humans, should live with that knowledge. If we are capable of stepping over that point, we will make progress in learning the new powers, which are there, behind that stage. The human body will heal itself, traveling to other worlds will be normal and our brain will leave the simple stage it is in now.

1-13
The source of everything

The source is not a starting point, because there is no starting point in the Universe, as I have said several times already. Time and distance are the limitations that we humans work with and believe in. Why do we do that? There is only one answer. Some people are afraid to lose face by confessing: " I WAS WRONG". Theories will be overthrown and will no longer be needed. Such people can be found in churches, science and politics. The system cannot admit its own lies! As we know, there is a lot of energy and most of it we cannot explain, not to mention work with. Some of it we do work with, such as electricity, radio waves, atoms and other simple energy fields. But let us try to analyze energy and where to find it. Energy is a field of waves that transport power to other places. Waves going to the other end of the line, to switch the light on, or to start our computer and keep it working. This is the supply for all those machines. We try to find an explanation in everything around us. The source, made of energy, is in everything we see. We need to believe that. We see a plant as a plant, an animal as an animal, a human as a human, a PC as a PC, and so on. All of these are made of energy. Because we have these energy limitations, we cannot imagine that we are only talking about energy, which is no more than waves. In the Matrix movie, we could see some simple explanations. In that movie, we were shown that nothing is real and that everything can be manipulated.

1-14
The Core

The Core, as I told you, is the complete energy of all that is going on around us. Without time, limitations and distance, the Core is near me and I feel it every second. The whole Universe is energy without limitations and we cannot manage this energy with our human brain. A human brain is so small, so simple and sometimes gets that extra energy from the soul world that is in connection with the source.

1-15
The start

They can create a human race that is capable of going deeper and getting on in life. The next step will certainly come. It is a matter of human time. I call this human time, because for the Universe and Core (energy) there is no time, no limitation. There is nothing that is against them. Maybe this small start by some scientists who are now talking about a field, is the start of a positive turn.

1-16
Explain energy

Let me try to explain. Everything is energy, and it expands regularly. It was the time when the Universe created the stars, planets and other places. Energy was expanding on a level you can see and not only feel it. The objects we can see, feel and touch are growing. Because energy is capable of creating everything that is needed or is good for the complete Universe. Creating is no problem, because there is no time and there are no limitations! Then it starts life, life we know in plants, trees and also human beings. These "living forms" are more complicated than the objects created before. And so we are at the level we are here right now. But… yes, there is another "but". With the creation of all living objects, these are also going more and more their own way. You can see it in the world of animals, plants and certainly in the human race. In the beginning, we were created of pure energy. In that way, we were connected with that energy, because it was created from the basis. A perfect body, made to survive in that world, in that time line. What we call "evolution" has been going on for millions of years and has changed our body, our brain and also our way of living. When we go back to the time of the Egyptians, Mayas and other great cultures, we can see things there that are signs from the source. Signs from that world, before the changes took place. Things were created in mysteries, inexplicable; nobody knows what happened there and why. We can see buildings, which will stand forever as long as needed. That was the start and at the same time, unfortunately, the end of the time with our energy source. Then people started to move away. Long before these cultures and the signs of the source, people were a closer part of that source. We can find several signs of rituals performed by these people at that time. The more you listen, read or feel, the more you will see one thing: We are now moving away from the source and the energy field simple because we believe in time and distance.

1-17
Time is not an issue

The energy that we come from, is just an expression of that energy and in the beginning we were "pure" human bodies that were directly con-nected with the energy. In some old buildings around the world, we can find evidence that people knew how to fly, how to build planes and so on. It was not a problem, because they were so pure that they knew how to look back in time and into the future. They knew what was there and what was coming. Time was not an issue, neither for them in those day nor for some people today. Some people? Yes, actually there are people who can see and know what was there and what is coming. BUT, the problem is that the people today are more dominated by their brain and so their way of seeing is no longer pure. The connection with the energy field is fading away and so it is so much more difficult to get a pure connection with that field. Some believe that they reach a higher level through meditation, that they can travel and see what is going on. There are people who can see others who are as far as them, so they can travel in time together. I hope I will remain levelheaded on this earth, keeping the connection I have now with this energy and all its powers.

1-18
No time

Plant, that rock, that water and everything you see around you. If you can reach the level of energy, there is no time, no past, no present and no future, because all of those were once and are already done. That is why some people can see into the future and do extraordinary things. In the world of energy, there are no limits and so you can be one with the Core.

1-19
No beginning no end

All scientists who reached that point of pure energy, people who have seen and known the key, are locked up with the label "mad". And so it will continue to go in the future until the truth will come out. Under the pyramids in Egypt, there is proof locked away. The same is true in the USA and other places where there is proof that energy builds everything. The powers that we are trying to get access to this information in these places, but the way these people worked with this information will never come out. Do they really think to find the key to that absolute energy in buildings of thousands of years old? No, those people also tried to find the key. They were more advanced than we are now, because they were more connected with the energy, but they did not know how to explain. It is incredible that we human beings always look back to what others have done. Trying to find a way in science and mathematics, people keep going the wrong way all the time they are here. Older generations knew how to work with heavy and strange materials, because they got the power to lift tons of rocks without touching, to create buildings that we know they built with a purpose. There will never be an explanation, because it was never written and it is not important to know for the present time. But why keep searching if

- First of all, there is no time.

- Secondly, there is no beginning and no end.

- Thirdly, because there is no time, every situation can be filled in, when you want and where you want already know? It seems to be human nature to make everything as complicated as possible and get things written incorrectly, so that nobody understands afterwards what it meant. It is foolish to believe that we should go back in time and see what they did. We know now what to do and that is get back to the Core, the energy I know will be the key and the answer to all our questions.

1-20
What is the end?

The end, something that is definite. We talk about "The end" in different ways. The end of the day, the end of a movie, the end of everything, or the end of our life. There are both old and new cultures that see "The end" as a beginning. Others see it like going to another place, and there are people who believe that they go into another dimension. But most people see "The end" as an absolute end. Absolutely the end! There is NO end and there is NO beginning. But how can I explain that? Every time we think "This is the end", it is just a part that is closed, not closing the way we think. Let me try to explain it this way: closing a part I compare to going to school, getting information from a teacher. After we have received that information, it is up to us to do something with it. Having received the information and then working with it, we show that we have learned something. In this case, we do not see that part of life as an end, but as a learning process. In life there are so many lessons to learn and it goes on forever; we learn something new all the time. But strangely enough, when the lesson is learned, we say "this is the end". We see it as closing part of our life. That is the way we see it and it is strange that, when there are certain periods in life, we believe that we have to close them first and then go on. We should regard them not as jobs being done, but as lessons learned. We have to work with those lessons for the rest of the time here on earth. There is no closing. This may be a strange idea, but that's the way it is. I see it like this: In my life - which is anything but boring - there is something happening all the time. Every day, there are situations when I think "How can I manage this?" And every time, I find out what is happening and how I should proceed. I have closed that part and I went on, is what most people would think. My mind and my connection with the Universe tell me something different.

- First of all, there is no time.

- Secondly, there is no beginning and no end.

- Thirdly, because there is no time, every situation can be filled in, when you want and where you want.

1-21
The human end

Pfffffffff, you must think that I have reached the end! Just as well that I do not believe in an end! Time is only there, because people adopted the concept, when they decided to start a life on this planet. The limitations of time are inherited from people here on earth. It is incredible how many limitations people believe in. Because they believe that those limitations are there, they exist! It's we who believe in time and all its limitations. Which brings me to my second point, which is believing in a beginning and an end. These limitations of beginning and end were created by human beings. I'm sure you can follow me in this. Now the third point: Fill in your situation and the time you want! This will be the hardest part to explain. If you have come to the point that you are above the limitations of time, you will see you can live your life any way you want. All these situations, these parts, you can place behind each other the way you want. It is not a puzzle that has only one way to finish. A human life is not lived that way. Every second, we end up in different situations and then we get action and reaction. All these apparently insignificant situations together make up something important for us to reach a higher state. We get input from outside and reach a point where larger moves manifest themselves. Having completing these situations, newer cycles follow and they teach you how to manage life. Life is like a continuous learning experience, not a place with beginnings and ends.

Look at how life goes. We are born full of energy, clean and pure. We follow the way of an earthly life and we become weaker and weaker, believing we are growing, but that is a misconception. We do not grow. On the contrary, all the situations we end up in, we see our body breaking down to a weak old collection of bones and flesh. That body, which is there for us, ready to help us and willing to go the way we want to. We abuse it, until it is empty and can no longer fill in our needs. That is the point when our mind starts to break down as well, because it cannot handle all the inputs and needs of true life. It's as if our body and mind have reached point of feeling useless, painful and old. What is happening? Because we believe in a world of time, we act accordingly. When we are young and strong, we can do everything. Getting older, we turn weak and full of pain. We do extreme things with our body. Things a body does not need to do, but we act as if we believe we have to break down that piece of flesh! There is something strange going in that. It's as if we are all here, because we are in the middle of an experiment.

1-22
What is the human body capable of?

"What is the human body capable of." Is that the reason why we are here? I do not think so, because this is what you also create in your own life. Step by step, you are creating your own "end", because you want it that way! Is not one of your main goals to create a good life for the future, gathering enough money for later, when you are older? Is not that like living with the knowledge of getting old and being weak later? People die, because they believe in that end. Stop believing in time and follow the path that is open for you. If we can see further, we reach a point where we can see that we are no more and no less than energy. Pure energy that is capable of doing anything. We are guests in our bodies, which are not ours and only there to learn from. People who believe in energy, act like hosts for their bodies and do everything to show it the way in a beautiful life. It is hard to believe in energy when you are in pain or in a difficult situation. Once you find the connection with the energy again, there are no more limitations, because you are in a world which has no beginning and no end. We borrow our bodies for as long as we need them, for the lessons we need to learn. After that, the body goes back to energy and returns for other energy's need and help. As long as we see ourselves and our life just as being here on earth for some years, there is no learning, no real path you are following. "The end" is created by human beings and it will continue to be so as long as people believe in it. Follow the path of energy, it is wonderful and it gives you everything you can imagine. Do not believe in these earthly matters any more, such as pain and domination. Then energy will be released that gives you the power you need. It is not your mind, it is not another way of thinking, it is energy that is talking through your soul to your hearth. Energy will guide you through this lesson that we call "Human Life".

1-23
The world beyond absolute zero

As we all know, mankind is trying hard to reach absolute zero. A point where we will find a start, a beginning or a solution. Do you really think that this point will tell us everything? A single molecule, a small part of what you think you see. A single atom, which is alone in this big world. That one atom will tell us.......... What do you think you will find? How earth was created? How life started? How we came into this life? Answers that we think we will get if we have gained possession of that one atom. But then the moment comes when we get the information of what that one atom really implies. And what do we get? We see new life forms, a new world in this single atom, in the absolute zero, where we thought we would find all answers! We see another world, a new beginning and a new life form. So absolute zero is nothing more than the door to a new world. That must be frustrating for all scientists who have been working hard for many years to get to that point. The point they dreamt of. "If we get there, we will have the solutions." Absolute zero, an entry point into another world and another dimension. It is certainly a dimension, because zero is, for us, an absolute. In the Universe, it is nothing more than "another step". In the world of matter we think, live and act like matter, because that is what we have learned. In the world of the Universe, there is no beginning, no end, no time and no matter. Everything is made of energy, and there is no zero. We have created that zero, over and over, and we continually set the point higher or lower, where we think zero is supposed to be. Zero = nothing. Do you believe in nothing? Nothing does not exist, because the nothing is made by something; something we call energy in the Universe. Even if I get zero from you, I can feel you and I get your energy. Because everything in life is interconnected. Everything has something to do with some other part in life, other people and other places. Why do we have people around us? Because we like them? Because we need them? Do we have to do something for them? No, because they are a part of our energy field and so they will come and go, during our life. In that way, they participate in Universal life. The way we react when people are around us, provides the Universe with the energy it needs. The connections between people, negative or positive, are made of energy and that is what we and the Universe are there for. But why do we have these negative people around us? Because of the failing system perhaps? Or could it be a change of the human race? Is the world changing? All of these are parts of the same thing, because the human race needs negative input to find the right positive output! If we only live like we are in paradise, happy, with healthy food and a perfect place around us, then what will this mean for the energy? This energy will be stable, but not growing. The environment we live in, is a place where negative energy is needed, so we can take positive energy to a higher level. Every second, this energy goes to another level. Some people believe the next level will come soon. I believe the next level is there for the first ones among us and that there are already others who have reached higher levels. In the energy world, level is not on a specific place, but level is the energy you can feel and master and know what to do with. That is why there are at this moment those extreme differences between people. Some are violent criminals and believe in war. Others believe in a world of understanding and peace. Both kinds of people need each other. I know that it looks like the criminals own the world. That is because they get all the attention in the media. In the background, there are those who work hard to reach the next level in their lives. Some people believe in the "new" indigo children and we should respect them. I wrote a couple of chapters on this subject and I believe that these human beings have always been there. Now is the time they have to come out. They have the strength to take us further. For a long time, we have thought that fighting is the answer to everything, because we had adopted the instinct of animals, but even animals are not as cruel as we are. Human beings have dropped below the level of animals, and now we should grow again. We should start to get back to the level we were at thousands of years ago. When we have reached that level again, we are entering the world of zero. Let's return to energy and absolute zero. As you know, in the Universe there is no time, so no beginning and no end. This is because everything is turning around in the world of energy. Man once thought that the world was flat and we would fall from it if we kept walking. We now know that the world is round. We believed that we could not go into space unlimitedly, because the distances are too great for one human life. If you see what I see, you will be silent. All it takes is to believe that there is no time and no distance. Why do we always impose limitations on ourselves, believing there are things that we cannot do? This is also what happens with this "absolute zero". Why do we create a point of limitation? The explanation is easy: because we think it will give us the power and the explanation of the creation of the world! The negative energy we put into this is so strong, so extreme. We will be very surprised, in a disappointing way, when we reach that point. Working with energy in fields in which you never believed, is dangerous. I am not a scientist and I cannot tell you mathematical solutions, nor can I prove new theories. It is not up to me to do that. But I can tell you that there are no answers in that point called "zero", only a new world. People believe that this zero is interesting, because it would enable them to change the world, dominate one atom and build perfect machines. This atom is part of the energy field as a whole and we are also parts of that same field. This atom is a new world and the minute we reach the point where we separate this atom from the energy field, we will enter a world we cannot understand and do not know how to work with. A world that is no longer pure matter, but an entrance to a new world of energy. It is certainly interesting but we cannot handle it with machines. None of the machines we have, none of the knowledge we think we possess will tell us what we see and what we can do with it. It will pose a new problem for scientists. "What to do now?" This level can no longer be handled with our machines. They hope to find in the world of the quantum theory, a solution to go further. Instead, they go deeper into a new world and a new science. This science should be built in the world of energy. It is true that there are already people who have started this work. As far as I can see, these people merely wanted to get publicity. They like to manipulate their power. They will not get us any further, as far as I can see. In the world of energy, there will never be an explanation when you enter that field with these intentions. Energy feels everything, sees everything and is pure. As long as we practice science to get more power, we will get stuck in never-ending worlds. Those worlds are easy to create, because energy is very flexible, but above all, it is pure. So, forget about absolute zero, it does not exist! Absolute zero is another point that mankind has created to believe in. I went beyond that point and you can go there too if you want. It is the same place where you can enter another dimension. Now I suppose you think that I am going too far. I admit that I believe in things that are perhaps difficult to grasp. If this book still exists in 50 years' time, I know it will be read as something normal. There will be a time that this way of thinking and this way of living has become part of every life. Believe now what you want, take out of it what you need, or even do not believe in it. I know this is a wake-up call for you to get the energy you need, later in life. It is not my intention to push you onto a path you cannot believe in. It is a path that has to open up in your life. Nobody can help you to find that gate. The same gate we will find when we reach the absolute zero, where a new dimension presents itself.

1-24
Time does not exist!

That is right, but many people do not know that. The only thing they experience is that after a while, every problem is solved. Not always in the way you expect, but the right solution will always come. The wisdom that older people have, is very important for us and we need it in life. Older people have lived their lives for many years. They had lots of problems to solve and they experienced many good and bad things. But one thing is that they can see that every time there will be a solution for any problem. Time is the solution to everything. We, younger people, think we can manipulate time.

We think we can rush things. That's why most problems take so long to be solved. After a lifetime, the problem is probably still there, because we tried to manipulate the situation. Older people have found out that taking that one moment and live that second, solves everything.

1-25
In a world of no time

There is no far away or deep inside, everything is near us. The minute we can see that, we have reached the point where we can go further, or in my terminology "going deeper in life".

1-26
So far

So far, I would like to leave aside for a moment the quotes from my earlier book in order to continue on the other side of the world. We know that there is "more" going on around us than what we hear but also what we feel and see. The parallel worlds are not figments of our imagination, just like the many dimensions that are all aligned. By interpreting time and clinging to an artificial whole, today's science hovers between denial and proof. This brings us to a chapter that is completely obscure and little information is available.

$$E = mc^2$$

E = energy

m = mass

c = the speed of light

Chapter 2

Science first

2-1
Dark matter, a dark corner

We are now going to pry into a dark corner and come across remarkable data. By first letting science have its say, we clearly see that this corner of the 'corner of knowledge' bases everything on assumptions as there is hardly any evidence, if any at all! Worse still, they do not know what they are talking about and they just make up a story around it.

So science first.

2-1 a
Dark matter

Dark matter is a form of matter thought to account for approximately 85% of the matter in the universe and about 27% of its total mass–energy density or about $2.241 \times 10{-}27$ kg/m3. Its presence is implied in a variety of astrophysical observations, including gravitational effects that cannot be explained by accepted theories of gravity unless more matter is present than can be seen. For this reason, most experts think that dark matter is abundant in the universe and that it has had a strong influence on its structure and evolution. Dark matter is called dark because it does not appear to interact with the electromagnetic field, which means it does not absorb, reflect or emit electromagnetic radiation, and is therefore difficult to detect.

Primary evidence for dark matter comes from calculations showing that many galaxies would fly apart, or that they would not have formed or would not move as they do, if they did not contain a large amount of unseen matter.

Other lines of evidence include observations in gravitational lensing and in the cosmic microwave background, along with astronomical observations of the observable universe's current structure, the formation and evolution of galaxies, mass location during galactic collisions, and the motion of galaxies within galaxy clusters. In the standard Lambda-CDM model of cosmology, the total mass–energy of the universe contains 5% ordinary matter and energy, 27% dark matter and 68% of a form of energy known as dark energy. Thus, dark matter constitutes 85% of total mass, while dark energy plus dark matter constitute 95% of total mass–energy content.

Because dark matter has not yet been observed directly, if it exists, it must barely interact with ordinary baryonic matter and radiation, except through gravity. Most dark matter is thought to be non-baryonic in nature; it may be composed of some as-yet undiscovered subatomic particles. The primary candidate for dark matter is some new kind of elementary particle that has not yet been discovered, in particular, weakly interacting massive particles (WIMPs). Many experiments to directly detect and study dark matter particles are being actively undertaken, but none have yet succeeded. Dark matter is classified as "cold", "warm", or "hot" according to its velocity (more precisely, its free streaming length). Current models favor a cold dark matter scenario, in which structures emerge by gradual accumulation of particles.

Although the existence of dark matter is generally accepted by the scientific community, some astrophysicists, intrigued by certain observations which do not fit some dark matter theories, argue for various modifications of the standard laws of general relativity, such as modified Newtonian dynamics, tensor–vector–scalar gravity, or entropic gravity. These models attempt to account for all observations without invoking supplemental non-baryonic matter.

2-1 b
History

Early history

The hypothesis of dark matter has an elaborate history. In a talk given in 1884, Lord Kelvin estimated the number of dark bodies in the Milky Way from the observed velocity dispersion of the stars orbiting around the center of the galaxy. By using these measurements, he estimated the mass of the galaxy, which he determined is different from the mass of visible stars. Lord Kelvin thus concluded "many of our stars, perhaps a great majority of them, may be dark bodies". In 1906 Henri Poincaré in "The Milky Way and Theory of Gases" used "dark matter", or "matière obscure" in French, in discussing Kelvin's work.

The first to suggest the existence of dark matter using stellar velocities was Dutch astronomer Jacobus Kapteyn in 1922. Fellow Dutchman and radio astronomy pioneer Jan Oort also hypothesized the existence of dark matter in 1932. Oort was studying stellar motions in the local galactic neighborhood and found the mass in the galactic plane must be greater than what was observed, but this measurement was later determined to be erroneous.

In 1933, Swiss astrophysicist Fritz Zwicky, who studied galaxy clusters while working at the California Institute of Technology, made a similar inference. Zwicky applied the virial theorem to the Coma Cluster and obtained evidence of unseen mass he called dunkle Materie ('dark matter'). Zwicky estimated its mass based on the motions of galaxies near its edge and compared that to an estimate based on its brightness and number of galaxies. He estimated the cluster had about 400 times more mass than was visually observable. The gravity effect of the visible galaxies was far too small for such fast orbits, thus mass must be hidden from view. Based on these conclusions, Zwicky inferred some unseen matter provided the mass and associated gravitation attraction to hold the cluster together. Zwicky's estimates were off by more than an order of magnitude, mainly due to an obsolete value of the Hubble constant; the same calculation today shows a smaller fraction, using greater values for luminous mass. Nonetheless, Zwicky did correctly conclude from his calculation that the bulk of the matter was dark.

Further indications the mass-to-light ratio was not unity came from measurements of galaxy rotation curves. In 1939, Horace W. Babcock reported the rotation curve for the Andromeda nebula (known now as the Andromeda Galaxy), which suggested the mass-to-luminosity ratio increases radially. He attributed it to either light absorption within the galaxy or modified dynamics in the outer portions of the spiral and not to the missing matter he had uncovered. Following Babcock's 1939 report of unexpectedly rapid rotation in the outskirts of the Andromeda galaxy and a mass-to-light ratio of 50; in 1940 Jan Oort discovered and wrote about the large non-visible halo of NGC 3115.

1970s

Vera Rubin, Kent Ford, and Ken Freeman's work in the 1960s and 1970s provided further strong evidence, also using galaxy rotation curves. Rubin and Ford worked with a new spectrograph to measure the velocity curve of edge-on spiral galaxies with greater accuracy. This result was confirmed in 1978. An influential paper presented Rubin and Ford's results in 1980. They showed most galaxies must contain about six times as much dark as visible mass; thus, by around 1980 the apparent need for dark matter was widely recognized as a major unsolved problem in astronomy.

At the same time Rubin and Ford were exploring optical rotation curves, radio astronomers were making use of new radio telescopes to map the 21 cm line of atomic hydrogen in nearby galaxies. The radial distribution of interstellar atomic hydrogen (H-I) often extends to much larger galactic radii than those accessible by optical studies, extending the sampling of rotation curves – and thus of the total mass distribution – to a new dynamical regime. Early mapping of Andromeda with the 300 foot telescope at

Green Bank and the 250 foot dish at Jodrell Bank already showed the H-I rotation curve did not trace the expected Keplerian decline. As more sensitive receivers became available, Morton Roberts and Robert Whitehurst were able to trace the rotational velocity of Andromeda to 30 kpc, much beyond the optical measurements. Illustrating the advantage of tracing the gas disk at large radii, Figure 16 of that paper combines the optical data (the cluster of points at radii of less than 15 kpc with a single point further out) with the H-I data between 20–30 kpc, exhibiting the flatness of the outer galaxy rotation curve; the solid curve peaking at the center is the optical surface density, while the other curve shows the cumulative mass, still rising linearly at the outermost measurement. In parallel, the use of interferometric arrays for extragalactic H-I spectroscopy was being developed. In 1972, David Rogstad and Seth Shostak published H-I rotation curves of five spirals mapped with the Owens Valley interferometer; the rotation curves of all five were very flat, suggesting very large values of mass-to-light ratio in the outer parts of their extended H-I disks.

A stream of observations in the 1980s supported the presence of dark matter, including gravitational lensing of background objects by galaxy clusters, the temperature distribution of hot gas in galaxies and clusters, and the pattern of anisotropies in the cosmic microwave background. According to consensus among cosmologists, dark matter is composed primarily of a not yet characterized type of subatomic particle. The search for this particle, by a variety of means, is one of the major efforts in particle physics.

2-1 c
Technical definition

In standard cosmology, matter is anything whose energy density scales with the inverse cube of the scale factor, i.e., $\rho \times a-3$. This is in contrast to radiation, which scales as the inverse fourth power of the scale factor $\rho \times a-4$, and a cosmological constant, which is independent of a. These scalings can be understood intuitively: For an ordinary particle in a cubical box, doubling the length of the sides of the box decreases the density (and hence energy density) by a factor of 8 (= 23). For radiation, the energy density decreases by a factor of 16 (= 24), because any act whose effect increases the scale factor must also cause a proportional redshift. A cosmological constant, as an intrinsic property of space, has a constant energy density regardless of the volume under consideration.

In principle, "dark matter" means all components of the universe which are not visible but still obey ρ x a−3. In practice, the term "dark matter" is often used to mean only the non-baryonic component of dark matter, i.e., excluding "missing baryons." Context will usually indicate which meaning is intended.

2-2
Observational evidence

2-2 a
Galaxy rotation curves

Rotation curve of a typical spiral galaxy: predicted (A) and observed (B). Dark matter can explain the 'flat' appearance of the velocity curve out to a large radius.

The arms of spiral galaxies rotate around the galactic center. The luminous mass density of a spiral galaxy decreases as one goes from the center to the outskirts. If luminous mass were all the matter, then we can model the galaxy as a point mass in the centre and test masses orbiting around it, similar to the Solar System. From Kepler's Second Law, it is expected that the rotation velocities will decrease with distance from the center, similar to the Solar System. This is not observed. Instead, the galaxy rotation curve remains flat as distance from the center increases.

If Kepler's laws are correct, then the obvious way to resolve this discrepancy is to conclude the mass distribution in spiral galaxies is not similar to that of the Solar System. In particular, there is a lot of non-luminous matter (dark matter) in the outskirts of the galaxy.

2-2 b
Velocity dispersions

Stars in bound systems must obey the virial theorem. The theorem, together with the measured velocity distribution, can be used to measure the mass distribution in a bound system, such as elliptical galaxies or globular clusters. With some exceptions, velocity dispersion estimates of elliptical galaxies do not match the predicted velocity dispersion from the observed mass distribution, even assuming complicated distributions of stellar orbits.

As with galaxy rotation curves, the obvious way to resolve the discrepancy is to postulate the existence of non-luminous matter.

2-2 c
Galaxy clusters

Galaxy clusters are particularly important for dark matter studies since their masses can be estimated in three independent ways:

From the scatter in radial velocities of the galaxies within clusters
From X-rays emitted by hot gas in the clusters. From the X-ray energy spectrum and flux, the gas temperature and density can be estimated, hence giving the pressure; assuming pressure and gravity balance determines the cluster's mass profile.

Gravitational lensing (usually of more distant galaxies) can measure cluster masses without relying on observations of dynamics (e.g., velocity). Generally, these three methods are in reasonable agreement that dark matter outweighs visible matter by approximately 5 to 1.

2-2 d
Gravitational lensing

One of the consequences of general relativity is massive objects (such as a cluster of galaxies) lying between a more distant source (such as a quasar) and an observer should act as a lens to bend the light from this source. The more massive an object, the more lensing is observed.

Strong lensing is the observed distortion of background galaxies into arcs when their light passes through such a gravitational lens. It has been observed around many distant clusters including Abell 1689. By measuring the distortion geometry, the mass of the intervening cluster can be obtained. In the dozens of cases where this has been done, the mass-to-light ratios obtained correspond to the dynamical dark matter measurements of clusters. Lensing can lead to multiple copies of an image. By analyzing the distribution of multiple image copies, scientists have been able to deduce and map the distribution of dark matter around the MACS J0416.1-2403 galaxy cluster.

Weak gravitational lensing investigates minute distortions of galaxies, using statistical analyses from vast galaxy surveys. By examining the apparent shear deformation of the adjacent background galaxies, the mean distribution of dark matter can be characterized. The mass-to-light ratios correspond to dark matter densities predicted by other large-scale structure measurements. Dark matter does not bend light itself; mass (in this case the mass of the dark matter) bends spacetime. Light follows the curvature of spacetime, resulting in the lensing effect.

2-2 e
Cosmic microwave background

Although both dark matter and ordinary matter are matter, they do not behave in the same way. In particular, in the early universe, ordinary matter was ionized and interacted strongly with radiation via Thomson scattering. Dark matter does not interact directly with radiation, but it does affect the CMB by its gravitational potential (mainly on large scales), and by its effects on the density and velocity of ordinary matter. Ordinary and dark matter perturbations, therefore, evolve differently with time and leave different imprints on the cosmic microwave background (CMB).

The cosmic microwave background is very close to a perfect blackbody but contains very small temperature anisotropies of a few parts in 100,000. A sky map of anisotropies can be decomposed into an angular power spectrum, which is observed to contain a series of acoustic peaks at near-equal spacing but different heights. The series of peaks can be predicted for any assumed set of cosmological parameters by modern computer codes such as CMBFAST and CAMB, and matching theory to data, therefore, constrains cosmological parameters. The first peak mostly shows the density of baryonic matter, while the third peak relates mostly to the density of dark matter, measuring the density of matter and the density of atoms.

The CMB anisotropy was first discovered by COBE in 1992, though this had too coarse resolution to detect the acoustic peaks. After the discovery of the first acoustic peak by the balloon-borne BOOMERanG experiment in 2000, the power spectrum was precisely observed by WMAP in 2003–2012, and even more precisely by the Planck spacecraft in 2013–2015. The results support the Lambda-CDM model.

The observed CMB angular power spectrum provides powerful evidence in support of dark matter, as its precise structure is well fitted by the Lambda-

CDM model, but difficult to reproduce with any competing model such as modified Newtonian dynamics (MOND).

2-2 f
Structure formation

Structure formation refers to the period after the Big Bang when density perturbations collapsed to form stars, galaxies, and clusters. Prior to structure formation, the Friedmann solutions to general relativity describe a homogeneous universe. Later, small anisotropies gradually grew and condensed the homogeneous universe into stars, galaxies and larger structures. Ordinary matter is affected by radiation, which is the dominant element of the universe at very early times. As a result, its density perturbations are washed out and unable to condense into structure. If there were only ordinary matter in the universe, there would not have been enough time for density perturbations to grow into the galaxies and clusters currently seen.

Dark matter provides a solution to this problem because it is unaffected by radiation. Therefore, its density perturbations can grow first. The resulting gravitational potential acts as an attractive potential well for ordinary matter collapsing later, speeding up the structure formation process.

2-2 g
Bullet Cluster

If dark matter does not exist, then the next most likely explanation must be general relativity – the prevailing theory of gravity – is incorrect and should be modified. The Bullet Cluster, the result of a recent collision of two galaxy clusters, provides a challenge for modified gravity theories because its apparent center of mass is far displaced from the baryonic center of mass. Standard dark matter models can easily explain this observation, but modified gravity has a much harder time, especially since the observational evidence is model-independent.

2-2 h
Type Ia supernova distance measurements

Type Ia supernovae can be used as standard candles to measure extragalactic distances, which can in turn be used to measure how fast the universe has expanded in the past. Data indicates the universe is expanding at an accelerating rate, the cause of which is usually ascribed to dark energy. Since observations indicate the universe is almost flat, it is expected the total energy density of everything in the universe should sum to 1 (Ωtot $\approx$ 1). The measured dark energy density is $\Omega\Lambda \approx 0.690$; the observed ordinary (baryonic) matter energy density is $\Omega b \approx 0.0482$ and the energy density of radiation is negligible. This leaves a missing Ωdm ≈ 0.258 which nonetheless behaves like matter (see technical definition section above) – dark matter.

2-2 i
Sky surveys and baryon acoustic oscillations

Baryon acoustic oscillations (BAO) are fluctuations in the density of the visible baryonic matter (normal matter) of the universe on large scales. These are predicted to arise in the Lambda-CDM model due to acoustic oscillations in the photon–baryon fluid of the early universe, and can be observed in the cosmic microwave background angular power spectrum. BAOs set up a preferred length scale for baryons. As the dark matter and baryons clumped together after recombination, the effect is much weaker in the galaxy distribution in the nearby universe, but is detectable as a subtle

(≈1 percent) preference for pairs of galaxies to be separated by 147 Mpc, compared to those separated by 130–160 Mpc. This feature was predicted theoretically in the 1990s and then discovered in 2005, in two large galaxy redshift surveys, the Sloan Digital Sky Survey and the 2dF Galaxy Redshift Survey. Combining the CMB observations with BAO measurements from galaxy redshift surveys provides a precise estimate of the Hubble constant and the average matter density in the Universe. The results support the Lambda-CDM model.

2-2 j
Redshift-space distortions

Large galaxy redshift surveys may be used to make a three-dimensional map of the galaxy distribution. These maps are slightly distorted because distances are estimated from observed redshifts; the redshift contains a contribution from the galaxy's so-called peculiar velocity in addition to the dominant Hubble expansion term. On average, superclusters are expanding more slowly than the cosmic mean due to their gravity, while voids are expanding faster than average. In a redshift map, galaxies in front of a supercluster have excess radial velocities towards it and have redshifts slightly higher than their distance would imply, while galaxies behind the supercluster have redshifts slightly low for their distance. This effect causes superclusters to appear squashed in the radial direction, and likewise voids are stretched. Their angular positions are unaffected. This effect is not detectable for any one structure since the true shape is not known, but can be measured by averaging over many structures. It was predicted quantitatively by Nick Kaiser in 1987, and first decisively measured in 2001 by the 2dF Galaxy Redshift Survey. Results are in agreement with the Lambda-CDM model.

2-2 k
Lyman-alpha forest

In astronomical spectroscopy, the Lyman-alpha forest is the sum of the absorption lines arising from the Lyman-alpha transition of neutral hydrogen in the spectra of distant galaxies and quasars. Lyman-alpha forest observations can also constrain cosmological models. These constraints agree with those obtained from WMAP data.

2-3
Theoretical classifications

2-3 a
Composition

There are various hypotheses about what dark matter could consist of, as set out in the table below.

Dark matter can refer to any substance which interacts predominantly via gravity with visible matter (e.g., stars and planets). Hence in principle it need not be composed of a new type of fundamental particle but could, at least in part, be made up of standard baryonic matter, such as protons or neutrons. However, for the reasons outlined below, most scientists think the dark matter is dominated by a non-baryonic component, which is likely composed of a currently unknown fundamental particle (or similar exotic state).

2-3 b
Baryonic matter

Not to be confused with Missing baryon problem.

Baryons (protons and neutrons) make up ordinary stars and planets. However, baryonic matter also encompasses less common non-primordial black holes, neutron stars, faint old white dwarfs and brown dwarfs, collectively known as massive compact halo objects (MACHOs), which can be hard to detect.

However, multiple lines of evidence suggest the majority of dark matter is not made of baryons:

Sufficient diffuse, baryonic gas or dust would be visible when backlit by stars.

The theory of Big Bang nucleosynthesis predicts the observed abundance of the chemical elements. If there are more baryons, then there should also be more helium, lithium and heavier elements synthesized during the Big Bang. Agreement with observed abundances requires that baryonic matter makes up between 4–5% of the universe's critical density. In contrast, large-scale structure and other observations indicate that the total matter density is about 30% of the critical density.

Astronomical searches for gravitational microlensing in the Milky Way found at most only a small fraction of the dark matter may be in dark, compact, conventional objects (MACHOs, etc.); the excluded range of object masses is from half the Earth's mass up to 30 solar masses, which covers nearly all the plausible candidates.

Detailed analysis of the small irregularities (anisotropies) in the cosmic microwave background. Observations by WMAP and Planck indicate that around five-sixths of the total matter is in a form that interacts significantly with ordinary matter or photons only through gravitational effects.

2-3 c
Non-baryonic matter

Candidates for non-baryonic dark matter are hypothetical particles such as axions, sterile neutrinos, weakly interacting massive particles (WIMPs), gravitationally-interacting massive particles (GIMPs), supersymmetric particles, geons, or primordial black holes. The three neutrino types already observed are indeed abundant, and dark, and matter, but because their individual masses – however uncertain they may be – are almost certainly too tiny, they can only supply a small fraction of dark matter, due to limits derived from large-scale structure and high-redshift galaxies.

Unlike baryonic matter, nonbaryonic matter did not contribute to the formation of the elements in the early universe (Big Bang nucleosynthesis) and so its presence is revealed only via its gravitational effects, or weak lensing. In addition, if the particles of which it is composed are supersymmetric, they can undergo annihilation interactions with themselves, possibly resulting in observable by-products such as gamma rays and neutrinos (indirect detection).

Dark matter aggregation and dense dark matter objects

If dark matter is composed of weakly-interacting particles, an obvious question is whether it can form objects equivalent to planets, stars, or black holes. Historically, the answer has been it cannot, because of two factors:

It lacks an efficient means to lose energy

Ordinary matter forms dense objects because it has numerous ways to lose energy. Losing energy would be essential for object formation, because a particle that gains energy during compaction or falling "inward" under gravity, and cannot lose it any other way, will heat up and increase velocity and momentum. Dark matter appears to lack means to lose energy, simply because it is not capable of interacting strongly in other ways except through gravity. The virial theorem suggests that such a particle would not stay bound to the gradually forming object – as the object began to form and compact, the dark matter particles within it would speed up and tend to escape.

It lacks a range of interactions needed to form structures

Ordinary matter interacts in many different ways. This allows the matter to form more complex structures. For example, stars form through gravity, but the particles within them interact and can emit energy in the form of neutrinos and electromagnetic radiation through fusion when they become energetic enough. Protons and neutrons can bind via the strong interaction and then form atoms with electrons largely through electromagnetic interaction. But there is no evidence that dark matter is capable of such a wide variety of interactions, since it seems to only interact through gravity (and possibly through some means no stronger than the weak interaction, although until dark matter is better understood, this is only hopeful speculation).

In 2015–2017 the idea dense dark matter was composed of primordial black holes, made a comeback following results of gravitational wave measurements which detected the merger of intermediate mass black holes. Black holes with about 30 solar masses are not predicted to form by either stellar collapse (typically less than 15 solar masses) or by the merger of black holes in galactic centers (millions or billions of solar masses). It was proposed the intermediate mass black holes causing the detected merger formed in the hot dense early phase of the universe due to denser regions collapsing. A later survey of about a thousand supernovae detected no gravitational lensing events, when about eight would be expected if intermediate mass primordial black holes above a certain mass range accounted for the majority of dark matter.

The possibility atom-sized primordial black holes account for a significant fraction of dark matter was ruled out by measurements of positron and electron fluxes outside the Sun's heliosphere by the Voyager 1 spacecraft. Tiny black holes are theorized to emit Hawking radiation. However, the detected fluxes were too low and did not have the expected energy spectrum suggesting tiny primordial black holes are not widespread enough to account for dark matter. Nonetheless, research and theories proposing dense dark matter accounts for dark matter continue as of 2018, including approaches to dark matter cooling, and the question remains unsettled. In 2019, the lack of microlensing effects in the observation of Andromeda suggests tiny black holes do not exist.

However, there still exists a largely unconstrained mass range smaller than that can be limited by optical microlensing observations, where primordial black holes may account for all dark matter.

2-4
Free streaming length

Dark matter can be divided into cold, warm, and hot categories. These categories refer to velocity rather than an actual temperature, indicating how far corresponding objects moved due to random motions in the early universe, before they slowed due to cosmic expansion – this is an important distance called the free streaming length (FSL). Primordial density fluctuations smaller than this length get washed out as particles spread from overdense to underdense regions, while larger fluctuations are unaffected; therefore this length sets a minimum scale for later structure formation.

The categories are set with respect to the size of a protogalaxy (an object that later evolves into a dwarf galaxy): Dark matter particles are classified as cold, warm, or hot according to their FSL; much smaller (cold), similar to (warm), or much larger (hot) than a protogalaxy. Mixtures of the above are also possible: a theory of mixed dark matter was popular in the mid-1990s, but was rejected following the discovery of dark energy.

Cold dark matter leads to a bottom-up formation of structure with galaxies forming first and galaxy clusters at a latter stage, while hot dark matter would result in a top-down formation scenario with large matter aggregations forming early, later fragmenting into separate galaxies; the latter is excluded by high-redshift galaxy observations.

2-4 a
Fluctuation spectrum effects

These categories also correspond to fluctuation spectrum effects and the interval following the Big Bang at which each type became non-relativistic. Davis et al. wrote in 1985:

Candidate particles can be grouped into three categories on the basis of their effect on the fluctuation spectrum (Bond et al. 1983). If the dark matter is composed of abundant light particles which remain relativistic until shortly before recombination, then it may be termed "hot". The best candidate for hot dark matter is a neutrino ... A second possibility is for the dark matter particles to interact more weakly than neutrinos, to be less abundant, and to have a mass of order 1 keV. Such particles are termed "warm dark matter", because they have lower thermal velocities than massive neutrinos ... there are at present few candidate particles which fit this description. Gravitinos and photinos have been suggested (Pagels and Primack 1982; Bond, Szalay and Turner 1982) ... Any particles which became nonrelativistic very early, and so were able to diffuse a negligible distance, are termed "cold" dark matter (CDM). There are many candidates for CDM including supersymmetric particles.

2-4 b
Alternative definitions

Another approximate dividing line is warm dark matter became non-relativistic when the universe was approximately 1 year old and 1 millionth of its present size and in the radiation-dominated era (photons and neutrinos), with a photon temperature 2.7 million Kelvins. Standard physical cosmology gives the particle horizon size as 2 c t (speed of light multiplied by time) in the radiation-dominated era, thus 2 light-years. A region of this size would expand to 2 million light-years today (absent structure formation). The actual FSL is approximately 5 times the above length, since it continues to grow slowly as particle velocities decrease inversely with the scale factor after they become non-relativistic. In this example the FSL would correspond to 10 million light-years, or 3 megaparsecs, today, around the size containing an average large galaxy.

The 2.7 million K photon temperature gives a typical photon energy of 250 electronvolts, thereby setting a typical mass scale for warm dark matter: particles much more massive than this, such as GeV–TeV mass WIMPs, would become non-relativistic much earlier than one year after the Big Bang and thus have FSLs much smaller than a protogalaxy, making them cold. Conversely, much lighter particles, such as neutrinos with masses of only a few eV, have FSLs much larger than a protogalaxy, thus qualifying them as hot.

2-4 c
Cold dark matter

Cold dark matter offers the simplest explanation for most cosmological observations. It is dark matter composed of constituents with an FSL much smaller than a protogalaxy. This is the focus for dark matter research, as hot dark matter does not seem capable of supporting galaxy or galaxy cluster formation, and most particle candidates slowed early.

The constituents of cold dark matter are unknown. Possibilities range from large objects like MACHOs (such as black holes and Preon stars or RAMBOs (such as clusters of brown dwarfs), to new particles such as WIMPs and axions.

Studies of Big Bang nucleosynthesis and gravitational lensing convinced most cosmologist that MACHOs cannot make up more than a small fraction of dark matter. According to A. Peter: "... the only really plausible dark-matter candidates are new particles."

The 1997 DAMA/NaI experiment and its successor DAMA/LIBRA in 2013, claimed to directly detect dark matter particles passing through the Earth, but many researchers remain skeptical, as negative results from similar experiments seem incompatible with the DAMA results.

Many supersymmetric models offer dark matter candidates in the form of the WIMPy Lightest Supersymmetric Particle (LSP). Separately, heavy sterile neutrinos exist in non-supersymmetric extensions to the standard model which explain the small neutrino mass through the seesaw mechanism.

2-4 d
Warm dark matter

Warm dark matter comprises particles with an FSL comparable to the size of a protogalaxy. Predictions based on warm dark matter are similar to those for cold dark matter on large scales, but with less small-scale density perturbations. This reduces the predicted abundance of dwarf galaxies and may lead to lower density of dark matter in the central parts of large galaxies. Some researchers consider this a better fit to observations. A challenge for this model is the lack of particle candidates with the required mass ≈ 300 eV to 3000 eV.

No known particles can be categorized as warm dark matter. A postulated candidate is the sterile neutrino: A heavier, slower form of neutrino that does not interact through the weak force, unlike other neutrinos. Some modified gravity theories, such as scalar–tensor–vector gravity, require "warm" dark matter to make their equations work.

2-4 e
Hot dark matter

Hot dark matter consists of particles whose FSL is much larger than the size of a protogalaxy. The neutrino qualifies as such particle. They were discovered independently, long before the hunt for dark matter: they were postulated in 1930, and detected in 1956. Neutrinos' mass is less than $10-6$ that of an electron. Neutrinos interact with normal matter only via gravity and the weak force, making them difficult to detect (the weak force only works over a small distance, thus a neutrino triggers a weak force event only if it hits a nucleus head-on). This makes them 'weakly interacting light particles' (WILPs), as opposed to WIMPs.

The three known flavours of neutrinos are the electron, muon, and tau. Their masses are slightly different. Neutrinos oscillate among the flavours as they move. It is hard to determine an exact upper bound on the collective average mass of the three neutrinos (or for any of the three individually). For example, if the average neutrino mass were over 50 $eV/c2$ (less than $10-5$ of the mass of an electron), the universe would collapse. CMB data and other methods indicate that their average mass probably does not exceed 0.3 $eV/c2$. Thus, observed neutrinos cannot explain dark matter.

Because galaxy-size density fluctuations get washed out by free-streaming, hot dark matter implies the first objects that can form are huge supercluster-size pancakes, which then fragment into galaxies. Deep-field observations show instead that galaxies formed first, followed by clusters and super-clusters as galaxies clump together.

2-5
Detection of dark matter particles

If dark matter is made up of sub-atomic particles, then millions, possibly billions, of such particles must pass through every square centimeter of the Earth each second. Many experiments aim to test this hypothesis. Although WIMPs are popular search candidates, the Axion Dark Matter Experiment (ADMX) searches for axions. Another candidate is heavy hidden sector particles which only interact with ordinary matter via gravity.

These experiments can be divided into two classes: direct detection experiments, which search for the scattering of dark matter particles off atomic nuclei within a detector; and indirect detection, which look for the products of dark matter particle annihilations or decays.

2-5 a
Direct detection

Further information: Weakly interacting massive particles § Direct detection
Direct detection experiments aim to observe low-energy recoils (typically a few keVs) of nuclei induced by interactions with particles of dark matter, which (in theory) are passing through the Earth. After such a recoil the nucleus will emit energy in the form of scintillation light or phonons, as they pass through sensitive detection apparatus. To do this effectively, it

is crucial to maintain a low background, and so such experiments operate deep underground to reduce the interference from cosmic rays. Examples of underground laboratories with direct detection experiments include the Stawell mine, the Soudan mine, the SNOLAB underground laboratory at Sudbury, the Gran Sasso National Laboratory, the Canfranc Underground Laboratory, the Boulby Underground Laboratory, the Deep Underground Science and Engineering Laboratory and the China Jinping Underground Laboratory.

These experiments mostly use either cryogenic or noble liquid detector technologies. Cryogenic detectors operating at temperatures below 100 mK, detect the heat produced when a particle hits an atom in a crystal absorber such as germanium. Noble liquid detectors detect scintillation produced by a particle collision in liquid xenon or argon. Cryogenic detector experiments include: CDMS, CRESST, EDELWEISS, EURECA. Noble liquid experiments include ZEPLIN, XENON, DEAP, ArDM, WARP, DarkSide, PandaX, and LUX, the Large Underground Xenon experiment. Both of these techniques focus strongly on their ability to distinguish background particles (which predominantly scatter off electrons) from dark matter particles (that scatter off nuclei). Other experiments include SIMPLE and PICASSO.

Currently there has been no well-established claim of dark matter detection from a direct detection experiment, leading instead to strong upper limits on the mass and interaction cross section with nucleons of such dark matter particles. The DAMA/NaI and more recent DAMA/LIBRA experimental collaborations have detected an annual modulation in the rate of events in their detectors, which they claim is due to dark matter. This results from the expectation that as the Earth orbits the Sun, the velocity of the detector relative to the dark matter halo will vary by a small amount. This claim is so far unconfirmed and in contradiction with negative results from other experiments such as LUX, SuperCDMS and XENON100.

A special case of direct detection experiments covers those with directional sensitivity. This is a search strategy based on the motion of the Solar System around the Galactic Center. A low-pressure time projection chamber makes it possible to access information on recoiling tracks and constrain WIMP-nucleus kinematics. WIMPs coming from the direction in which the Sun travels (approximately towards Cygnus) may then be separated from background, which should be isotropic. Directional dark matter experiments include DMTPC, DRIFT, Newage and MIMAC.

2-5 b
Indirect detection

Indirect detection experiments search for the products of the self-annihilation or decay of dark matter particles in outer space. For example, in regions of high dark matter density (e.g., the centre of our galaxy) two dark matter particles could annihilate to produce gamma rays or Standard Model particle–antiparticle pairs. Alternatively, if a dark matter particle is unstable, it could decay into Standard Model (or other) particles. These processes could be detected indirectly through an excess of gamma rays, antiprotons or positrons emanating from high density regions in our galaxy or others. A major difficulty inherent in such searches is that various astrophysical sources can mimic the signal expected from dark matter, and so multiple signals are likely required for a conclusive discovery.

A few of the dark matter particles passing through the Sun or Earth may scatter off atoms and lose energy. Thus dark matter may accumulate at the center of these bodies, increasing the chance of collision/annihilation. This could produce a distinctive signal in the form of high-energy neutrinos. Such a signal would be strong indirect proof of WIMP dark matter. High-energy neutrino telescopes such as AMANDA, IceCube and ANTARES are searching for this signal. The detection by LIGO in September 2015 of gravitational waves, opens the possibility of observing dark matter in a new way, particularly if it is in the form of primordial black holes.

Many experimental searches have been undertaken to look for such emission from dark matter annihilation or decay, examples of which follow. The Energetic Gamma Ray Experiment Telescope observed more gamma rays in 2008 than expected from the Milky Way, but scientists concluded this was most likely due to incorrect estimation of the telescope's sensitivity.

The Fermi Gamma-ray Space Telescope is searching for similar gamma rays. In April 2012, an analysis of previously available data from its Large Area Telescope instrument produced statistical evidence of a 130 GeV signal in the gamma radiation coming from the center of the Milky Way. WIMP annihilation was seen as the most probable explanation.

At higher energies, ground-based gamma-ray telescopes have set limits on the annihilation of dark matter in dwarf spheroidal galaxies and in clusters of galaxies.

The PAMELA experiment (launched in 2006) detected excess positrons. They could be from dark matter annihilation or from pulsars. No excess antiprotons were observed.

In 2013 results from the Alpha Magnetic Spectrometer on the International Space Station indicated excess high-energy cosmic rays which could be due to dark matter annihilation.

2-5 c
Collider searches for dark matter

An alternative approach to the detection of dark matter particles in nature is to produce them in a laboratory. Experiments with the Large Hadron

Collider (LHC) may be able to detect dark matter particles produced in collisions of the LHC proton beams. Because a dark matter particle should have negligible interactions with normal visible matter, it may be detected indirectly as (large amounts of) missing energy and momentum that escape the detectors, provided other (non-negligible) collision products are detected. Constraints on dark matter also exist from the LEP experiment using a similar principle, but probing the interaction of dark matter particles with electrons rather than quarks. Any discovery from collider searches must be corroborated by discoveries in the indirect or direct detection sectors to prove that the particle discovered is, in fact, dark matter.

2-6
Alternative hypotheses

Further information: Alternatives to general relativity
Because dark matter has not yet been conclusively identified, many other hypotheses have emerged aiming to explain the observational phenomena that dark matter was conceived to explain. The most common method is to modify general relativity. General relativity is well-tested on solar system scales, but its validity on galactic or cosmological scales has not been well proven. A suitable modification to general relativity can conceivably eliminate the need for dark matter. The best-known theories of this class are MOND and its relativistic generalization tensor-vector-scalar gravity (TeVeS), f(R) gravity, negative mass, dark fluid, and entropic gravity. Alternative theories abound.

A problem with alternative hypotheses is that observational evidence for dark matter comes from so many independent approaches (see the "observational evidence" section above). Explaining any individual observation is possible but explaining all of them in the absence of dark matter is very difficult. Nonetheless, there have been some scattered successes for alternative hypotheses, such as a 2016 test of gravitational lensing in entropic gravity and a 2020 measurement of a unique MOND effect.

The prevailing opinion among most astrophysicists is that while modifications to general relativity can conceivably explain part of the observational evidence, there is probably enough data to conclude there must be some form of dark matter present in the Universe.

2-6 a
In popular culture

Mention of dark matter is made in works of fiction. In such cases, it is usually attributed extraordinary physical or magical properties. Such descriptions are often inconsistent with the hypothesized properties of dark matter in physics and cosmology.

2-7
Antimatter

According to physics, for every kind of elementary particle there is an antiparticle, some of whose physical properties are the same, but others (such as the electrical charge) are exactly the opposite.

Matter made up of antiparticles is called antimatter. When a particle collides with its antiparticle, they annihilate each other and their mass is converted into energy according to the mass-energy relation $E = mc^2$.

2-7 a
History

In 1928, Paul Dirac was the first to come up with a mathematical formulation for the electron in accordance with relativistic quantum mechanics (a combination of the special theory of relativity and quantum mechanics). His description predicted that the antiparticle of the electron should also exist. In 1932, the particle was discovered by Carl Anderson. He saw traces of a particle in a nebula that corresponded to an electron, but in a magnetic field symmetrically bent in the other direction: it had to be positively charged. He called this antiparticle of the electron positron.

2-7 b
Properties

An antiparticle has the same mass and spin as the corresponding particle. The other properties (namely the conserved quantities) are exactly the opposite. For example, the anti-electron (also called positron) has mass 0.511 MeV and spin $\frac{1}{2}$, just like the electron, but in contrast to that lepton number -1 and electric charge +1.

If a particle is unstable, so is its antiparticle, with the same average lifetime. The decay products of the antiparticle are then the antiparticles of the decay products of the particle.

The above examples are fermions, from which matter is built up. Bosons are different: these are interaction particles that do not belong to matter or antimatter. For example, the positive pion $\pi+$ and the negative $\pi-$ are each other's antiparticle, but they occur under the same conditions and there is no reason to call one a particle and the other an antiparticle. The neutral pion $\pi0$, the photon and many other neutral bosons are their own antiparticle; there too the term antimatter has little or no meaning.

Antimatter occurs on Earth under normal conditions. It is created in the laboratory or in a particle accelerator, or it is created in nuclear reactions. A PET- scan uses radioactive substances that emit a positron (anti-electron) as a result of a nuclear reaction. In 2002, CERN succeeded in combining antiprotons and positrons to form antihydrogen atoms and to study their properties. They turned out to behave just like ordinary hydrogen. At the end of 2010, 38 antihydrogen atoms were created and trapped for one tenth of a second. On 6 June 2011, CERN set a new record: it succeeded in holding 300 anti-hydrogen particles for more than 15 minutes.

2-7 c
Annihilation

When a particle and its antiparticle meet, annihilation can occur, a process in which both particles are destroyed and a lot of energy is released, according to $E=mc^2$.

In some cases, only the internal energy of the particles remains, which was manifest as their mass (it is said that the entire mass is converted into energy). This energy then escapes in the form of electromagnetic radiation. For example, an electron-positron pair decays into two photons. One gram of matter with one gram of antimatter, when fully annihilated, yields 1.8 × 1014 joules of energy, 43 kilotons of TNT, or the combustion energy of about 30,000 barrels of crude oil.

2-7 d
Antimatter in the Universe

In the part of the universe studied by man, there is almost only ordinary matter. This is remarkable given the above: from 'nothing', matter and antimatter would arise in equal quantities. There are various hypotheses about the cause.

During the Big Bang, approximately equal amounts of matter and antimatter were formed. However, there was slightly more matter than antimatter and after a large-scale annihilation process, only some matter remained. This is the reason why today's universe consists almost entirely of matter and why there is so much radiation in the universe (radiation originating from the annihilations).

Another hypothesis is that a large, distant (and as yet unobserved) part of the universe is composed entirely of antimatter. At the time of its formation, all matter would have ended up in one part and the antimatter in another. It should be noted that we observe celestial bodies primarily through the emission of photons, and a photon is equal to its antiparticle (a photon is the same as an antiphoton) so there is no difference to be seen.

In April 1997, it was discovered that positrons were formed in the centre of the galaxy. NASA's Compton Gamma Ray Observatory discovered clouds of positrons.

In July 2003, a team of researchers from NASA discovered that antimatter is formed during giant explosions on the sun, known as solar flares. The researchers used NASA's Reuven Ramaty High Energy Solar Spectroscopic Imager (RHESSI) to study the high-energy X-rays and gamma rays.

In 2011, the Alpha Magnetic Spectrometer was transported to the ISS with the last flight of space shuttle Endeavour. It will remain there for ten years, conducting research into antimatter and dark matter.

At a height of a few hundred kilometres, the Earth is surrounded by a belt in which there are relatively many antiprotons. These are captured by the Earth's magnetic field and remain in existence because there is little normal matter at this altitude, so that annihilation does not occur. Especially in the South Atlantic Anomaly, high concentrations are measured that cannot be explained by normal decay. This conclusion was reached by an international team of physicists in 2011 based on measurements from the PAMELA experiment (Payload for Antimatter Matter Exploration and Light-nuclei Astrophysics). This is a European satellite experiment that was launched five years ago. It is speculated that this antimatter could be used in an antimatter drive in the future.

2-8
Dark Energy

Dark energy is a hypothetical form of energy in the universe that would be responsible for accelerating the expansion of the universe. Dark energy is everywhere and evenly distributed in the universe. It behaves as if it were exerting negative gravity.

In 1917, Albert Einstein had already introduced a cosmological constant into his field equations. As Einstein assumed a static universe, he did this to prevent the universe from collapsing due to gravity, according to his theory. After the discovery of the expansion of the universe, Einstein withdrew the idea of this anti-gravity and called it "his greatest blunder".

Distribution of dark matter and dark energy in the universe relative to visible matter according to WMAP measurements

In the 1990s, the study of distant supernovae, the Supernova Cosmology Project, discovered that the expansion of the universe began to accelerate some five billion years after the Big Bang. The only way to explain this was to introduce an unknown force that behaved like a cosmological constant and acted like negative gravity.

Careful analysis of the WMAP data in March 2003 revealed that 74% of the total energy of the universe consists of dark energy. In the meantime, observations from the Planck Observatory have reduced this proportion to 68.3%. The mass energy of ordinary (baryonic) matter amounts to 4.9%, the remaining 26.8% is explained by dark matter. Cosmologists do not yet have an explanation for this dark energy.

They think about the energy of the vacuum itself, the so-called zero-point energy. However, this causes very big problems for the theorists when this energy is calculated according to quantum mechanics. The outcome is much higher (as much as 10120 to 10140 times) than the observed dark energy.

2-9
Remarkable discovery

Last Post March 2021

Physicists at CERN find a clue that could turn our understanding of reality upside down.
A brand new particle, a still unknown force of nature ... physicists at research institute CERN see hints of something that could turn our understanding of reality upside down.

'This result may mean that nature has an as yet unknown, fifth fundamental force of nature'.

The already known four fundamental forces of nature are:
the electromagnetic force,
the weak nuclear force,
the strong nuclear force,
the force of gravity.

I quote part of a 2015 report in Scientias.nl :

The Hungarian research group that previously provided the impetus for the Americans' suspicion of a fifth force of nature publishes another paper, endorsing the existence of the particle observed in 2015 - and thus the alleged fifth force of nature. Could it really be...?

Back to 2015

To understand exactly what the Hungarians saw and why some physicists get so excited about it, we have to go back to 2015. The year the Hungarians first reported the discovery of a particle that, on closer inspection, could well be the carrier of a fifth force of nature. "The Hungarian scientists have carried out a small experiment, in which they make 'impacted' beryllium nuclei by shooting protons at a lithium nucleus," explains Niels Tuning, a physicist at the National Institute for Subatomic Physics in the Netherlands (Nikhef). "After a short time, the impacted beryllium nucleus falls back to the ground state and emits a photon (light particle). Such a photon can split into an e- (electron) and an e+ (positron). The Hungarians have analysed these e+e pairs and there appear to be more e+e pairs with a large opening angle than expected. This may indicate the creation of a particle, which after decay also creates e+e pairs with a large opening angle. This is how we normally discover new particles: as a 'peak' in a distribution (normally this is the distribution of the so-called 'invariant mass' of a pair of particles, but in this paper they look at the distribution of the opening angle, which is strongly related to it)."

As mentioned, the Hungarians' paper, initially overshadowed by other news, lay in a cupboard gathering dust, until American researchers reinterpreted the results a year later. "They came to the conclusion that the (new particle the Hungarians had seen, ed.) could be a spin-1 particle. Photons, W + Z particles and gluons are also spin-1 particles, and are the particles responsible for the electromagnetic force, the weak nuclear force and the strong nuclear force respectively. Spin-1 particles are typically the particles that are exchanged and thus responsible for a fundamental force of nature." And in the case of the particle that the Hungarians had seen, it could therefore just be a carrier of a 'new' fundamental force of nature, according to the Americans.

While the Americans were fiddling with the Hungarians' results, the Hungarians themselves were not sitting still. Last month, they came up with a new paper that seems to verify the results from their first one. "In the first paper, they measured on 8Be (beryllium) nuclei and saw a 'peak' corresponding to a particle with a mass of 17 MeV, about 33 times heavier than an electron, but still very light compared to other particles we know," Tuning says. "In the new paper, they measured on a different nucleus, namely 4He (helium) and again measure a 'peak' at 17 MeV. So this corresponds well and is quite remarkable. Because the energy of the e+ and e- are slightly different, this corresponds to a slightly different opening angle (115 degrees instead of 140 degrees).

World-shattering

Whereas the Hungarians could count on little media attention in 2015, their new paper is definitely in the spotlight. After all, could they really be on to something fundamentally new? If follow-up experiments show that the Hungarians are right, it will be world news."It would be earth-shattering if there were more forces than the electromagnetic force, the weak nuclear force, the strong nuclear force and gravity," Tuning states. "It would undoubtedly win a Nobel Prize and take many studies in a new direction, both in accelerator particle physics and astroparticle physics, as well as cosmology." For example, the discovery of a fifth force of nature could have huge implications for fathoming one of the greatest mysteries in astronomy: dark matter. "Explaining dark matter requires particles that have mass (and therefore generate extra gravity), but they must not interact too much with ordinary particles to explain existing observations. This new X17 particle might be able to do that."

(Courtesy of Scientias.nl)

Postscript:

Here again we clearly see how mainstream science is shielding itself from everything because it is afraid that many things will change if such far-reaching discoveries are made. There was a lot of scepticism and even disbelief and "the chances are small that it is true". Now that CERN has made a similar measurement, a difficult situation has arisen. For what now, ladies and gentlemen scientists? It has always been so that what one does not know, one cannot see nor find! And yet science continues to protect these old-fashioned views.

2-10
Elements

We are now many pages ahead and what we see is that much has been written but little has actually been recorded, as there is hardly any data on black matter.

The most remarkable thing is the press release posted above with the latest data from CERN. There is a reference to a "fifth force of nature"! This is just the beginning of the new world that is being tapped into.

I myself have experienced many things from which I know that there is much more than what we feel, see and hear. We have the world that is placed under the "witch" world. A world that in reality is a world that certain people do see and experience.

In many writings it is written about:
-The world of bright light
-The world of fire

-The world of water
-The world of shadows
-The world of great cities in the clouds

Let us go through them all as I see and experience them.

2-10 a
The world of bright light

The world of bright light is quickly associated with the connection with goodness, heaven and the passing away of this form of life. But light is a source of life, a source of bright energy that nature needs in order to be what it is today.

People also want to see light as the pure thing, but also as the most powerful form when it comes to power and strength. I call that strange because the dark is many times stronger and has more mass. Those people who are working with dark matter know that too.

Light seems to be pure but in reality it is one of the most polluted forms we have around us. This will be incomprehensible for many and certainly if one is religiously sensitive with the world of angels and the belief in the good. Just as faith wants to make us believe and in itself does good work, because many follow it. But also those, who see it as a connection with the multiple worlds always show this as being light.

How far this goes, we see with the many organisations in a new world who think that energy is light! Here you come to the point that light is so lightweight in energy quality that many can carry it and believe in it. But believing in the many dark forces that are in the unknown dark or black is another thing. See how people are speculating about the so-called black holes and what we can experience there. See how people avoid black but also feel more comfortable in an illuminated space instead of a black cellar where you cannot see your hand before your eyes.

Afraid of the unknown but even more afraid of the many energies that have merged into a much stronger power! So intensely powerful that no light can come from it.

That makes light a fraction of the mighty black. That one does not search in the black is because it is still so inexplicable and there are few real studies about it.

2-10 b
The world of fire

The world of fire is the world usually associated with the world of hell and its devil. It is described in holy writings and is a way to keep mankind in the bonds of faith. Purely to impose things on people.

Naturally, the world of fire is a beginning of the new and the creation of new matter, often in the form of volcanoes and their eruptions. Fire is also connected to the inside of the earth, although that is still a big question, as more and more evidence is coming to light that the inside of the earth contains the primordial world and that the mantle is just lava and liquid.

This makes fire a strange player in our world, because as much as it creates new life, it also takes away so much life. What we know for sure about fire is that it is a power of the absolute and it brings together the present and the past. Fire is also the "beginning" of everything and clearly acts as an innovator.

2-10 c
The world of water

In the world of water we see that everything around us is ocean and there is no land where you could live in the human world. Yet you will be able to enter that world once you realise that you are not clinging to a human body which will only be in one place.

Water is the source of life, the source of our earthly existence. That is what we as humans think but also accept as a truth. But if we look deeper, we indeed see a seed germinating when it comes into contact with water and we see people living endlessly on just water. So the scientific conclusion is: water is the source of life.

Water in humans, on the other hand, shrouds everything in a dark curtain whose truth no one is allowed to see. And so we come to the curtain of water. For, what do we know about water and what is water? We now know a little more, because some scientists apply negative and positive energy to water and come up with some remarkable data. Water is an energy conductor, water is one of the ways of absorbing energy and converting it into whatever energy is needed at that moment in that place.

2-10 d
The world of shadows

In the world of shadows, people talk a lot about spirits or beings from an old world that are active at night.

When sleeping in a dark room, it is not uncommon to have visitors. In many cases it is not a pleasant experience, although in general these shadows do nothing else than stay around your bed.

In some cases, the shadow seeks contact and wants you to do something for him/her in the present world you are in. But in many cases, nothing will happen and neither if you make contact with their world. Remember, all life is on one line and it is a whole of energy that is released in many forms. Seeing the shadows can sometimes be frightening, especially since we have been taught that they are witches, aliens and dangerous demons. Now there are many sober people but a shadow has an effect even on those ones.

Think further, for if it is black energy, is it not your energy that has no place in your body?

We will certainly come back to that when we really start to see what our life is.

2-10 e
The world of big cities in the clouds

We have already seen some pictures of cities in clouds, but also cities that you see further away and assume they do not exist in the present world. Are they gateways to other dimensions? Seems unlikely, as there is no time, there is no past and present but only a now. The cities are there but they serve another world! So they are from another world, do not fit into the world you have to experience. They are not always intangible, but neither functional in the world in which today's man lives.

Seeing cities is a regular occurrence for me. For example, on the cover of this book you see our view of the sea, where very often on the right-hand side of the bay, I see big buildings. Sometimes when I sit on the balcony, it seems that we are connected to that city far out to sea. It is reality and it can be seen clearly.

In the same way I have experienced, twice, the strange objects above our house, charged with energy, and constantly generating lightning and strange clouds. I have already described this in detail a few times. I put it more in the form of extraterrestrial or earthly experiments. More and more I tend to believe that the distant city at sea is so close to our house that we can clearly see things and thus the energy.

I know that for a normal thinking person this is madness and unbelievable, but when we go into Dark matter, then things start to explain themselves.

Chapter 2

Chapter 3

Chapter 3

We have now really made the journey in time and we have also landed on several blank pages. These pages show what time, distance really is. On those blank pages you realise that everything that is written on this subject is too much!

A strange statement for a writer but you will certainly find the explanation in this book.

Too much written because we understand more and more that everything in today's world revolves around a clock, which lets you know to the nearest thousandth of a second how long you are working on something and again, for some, wastes valuable time. The entire current system is linked to this time.

3-1
- Time
- Distance
- Nothing

When I started to deal with the word time, I realised when reading the description that time is a human constraint.

The very first tool to indicate the time of day was the sundial, invented by the Ancient Egyptians and Mesopotamians. The oldest sundials were obelisks (3500 BC) and shadow clocks (1500 BC). ... Using the sun's shadow, they determined the times of day.

This was done purely to find a place in a day. So already then, 3500 years before Christ, these structures were made to indicate when something had to take place. Before that, man only had the indication; sun up and sun down.

By going deeper into the history of time, it became clear that it was a human "binding agent" to bring people in line. Now everything goes by a clock and people have become the slaves of time. An evolution of about 5000 years has made man slave to this fictitious fact.

Worse is that the complete science, until recently, linked everything to time and distance. Now, with the Quantum period, one has slowly but surely dropped the concept of time and distance, simply because it no longer fits into the new calculations and there are significant frictions between the outcome and the data.

Einstein also had a hard time with this at the end of his life. He realised that in many cases he had estimated time wrongly and that, as a result, formulas were no longer accurate. All his theories had been calculated using time and distance! He was shielded for in his last years of life he was described as "insane" and people did not really listen to him anymore! The reason was simple, because then the whole science would fall apart. Together with Einstien, other great scientists have also ended up in a spiral, where one needs heavy calculations to save some of the theories.

It is striking that great mathematicians but also many scientists who write books full of numbers and formulas, often at the end of their lives completely lose their way because they notice that errors have crept into their theses. Stephen Hawking (physicist, cosmologist) was such a man too, who spent his last years wandering more than he had come up with solid information.

Don't get me wrong, I don't want to criticise any of the great scientists. They are all humans who were pushed into a corner by large institutions and had to work with information that was wrong from the start. The basis of most calculations is already wrong, which is "time". All this is why science and its basis is corrupt and many things are formulated in such a way that everyone has to assume that it is correct.

The key in the whole thing is simplicity. Simply analysing and seeing what is happening, is the best way to go. What has happened for thousands of years; by putting science in an elite corner, where only selected approved persons for certain clubs, got the chance to publish theses. It's still going on now, but there has been a strong side branch since the Quantum showed its face, and there are groups that got to work on it.

Coming back to the "old" science, where scholars broke their heads about time machines, the past and the future. That there are connections with various worlds is nothing new, but that heavy, over-complicated machines had to be built was only because the word "money" was important. We see with Tesla, Rife and those people that simple devices perform miracles and theories come up that science is not so happy with. Now people are still busy trying to stop these things, for why still no free electricity and also the solution for tackling diseases! Everything is about money, a lot of money.

Science, pharmaceuticals and wars are money-grubbers, if the papers are to be believed. But does all the money go into the bank accounts of these mega-clubs?

The more I dig and the deeper I go, the more obvious it is that the books don't add up. A lot of money disappears and no, not in own pockets or foreign accounts but disappears into nothing! While writing the book "The long suppressed secret", I could not yet put my finger on the sore spot, namely that it was about the financing etc. of the "other civilisation". Hitler had indeed contributed a lot. But in order to finance things above ground for things under ground, such as the many tunnels and supply of goods above ground, money is needed! Yes, when I started to read financial reports and compare the many tax expenditures, a remarkable flow of money emerged worldwide. Controlled by the World Bank linked to all the small national banks, almost half of the entire world's capital disappears into thin air. No, we are not talking about a few billionaires or hyper-rich like the Jewish bankers! No, we are talking about numbers with many zeros circulating above ground to finance a second world, giving them all the resources they need.

The release of the Quantum is planned just like the whole system we are in, since the 21th century. It is a script that I have been writing about for many years, but unfortunately it is always dismissed as nonsense. Now with this whole virus madness, people are starting to see that the story stinks on all fronts and also that there are superior forces at work.

Who remembers the opening of CERN? Does anyone remember what the images were like?
So far, CERN is one of the most famous and expensive projects in the world, which has found, as it seems, the contact with the different worlds, in other words the there and here and the place with the fifth fundamental force of nature.

3-2
Time Machine

The time machine is often presented as a circle that is supposed to serve as a "portal". Through that portal and through different time settings you can go to certain places in a timeline and enter the past or future world. The doctrine of "Energenius life" obviously has a different theory about this and has proven that time travel takes place everywhere, no true gates are needed, but it's about "being open to the whole". However, there are also energy portals where one travels through time, but these portals were built in the past to go to certain energy sources, which were determined by the portalbuilders.

But we get stuck again with the question: How is that possible if there is no time?

I have described this in my book "Dimension and Hologram" and then you come to the shocking conclusion that time is on one line and it is not about going forward or backward in time.

How beautiful it is that, as I stated in "Energenius life", in Quantum Science one has also figured out that time is a disturbing factor instead of a place marker in life.

But back to the time machine, a device that interests many people and people still know too little about it. Each place where a time machine is found, is a conscious place of the entire world of energy. And I write this because people like to have something familiar around them. The place where the device is, or the "portal", is the point that you are locked by humanly imposed time. But suppose it is a line/entry with that one place in the whole world of energy. It is then time travel without a set point through this energy world.

It is odd that man has always started from a fixed point, and it is clearly a human acting and thinking, as man always thinks and acts from one point. In all the years of travelling in the world of energy, it is very clear that man can seldom oversee the whole thing, for everything is like a horse with blinkers on, so that man does not receive too much information!

If you travel to any place, you will find that there is a great unlimited energy and time represents a narrow location or a narrow way. That is why I often talk about the Quantiverse in true time travel, because it can go in any direction and there is no fixed base, place, date or time.

So, now we come to the question "you are writing this book in a certain time and currently in the year 2021, aren't you?"

True, if one sees it through the eyes of a human being who lives, works and lives according to the standards imposed by a system, which one has accepted unconditionally for centuries.In the true world of energy, there is no "fixed point" because this unknown source of energy has still not been discovered by man. Although data is slowly being released, it is still far from being understood by man or being able to place it in a human life. There is a lot going on in the whole and a lot of it cannot be placed somewhere because it is simply "the whole". You can compare it with the air around us. You feel it, you breathe it but it is not to be seen and not really tangible because it is literally everywhere!

In the world of energy it is even worse, because as human beings we are not really energy seers. We feel it, some more than others, but we do not see it and we hardly know what to do with it, since we think that energy is only in certain places. Believing that time exists, makes that sometimes we have distances with so many zeros that a piece of paper is not enough. In the world of energy, you can eliminate the zeros and shift instantly. All because we are that energy ourselves!

Now this seems to be a difficult concept and a difficult thing to work with in our daily system of thought, for how am I going to indicate at what time we are going to eat or be somewhere? And this is still a large impassable area for us to work on, as we do not and cannot yet see the true process of the whole thing. But we know from ancient writings, but also from people who have experienced a glimpse of this world, that much more is possible through the power of energy. But we know from old writings, but also from people who have experienced a glimpse of this world, that much more is possible through the power of energy. The precise of the old buildings, built without, what we know, High Tech equipment, but also the sometimes supernatural descriptions of things that are still often misinterpreted and thus not having the power.

Working with this power is not to be found in a book called "Energy for Dummies". It can be found in some books in a fragmentary way and then again as I wrote, the texts are put in the wrong context.

In the many Energenius journeys I have experienced, I have been amazed by how everything is possible. For example, it became clear to me that you create the world around you yourself and that you determine your own path and can deal with things that you think are far beyond your ability. But every living being has the ability to make of their life what he thinks it is! OK, I can hear you mumbling again "But not to me". The excuse of being born to poor parents or in a bad neighbourhood, is an excuse that makes no sense at all. What stops you from seeing another neighbourhood as your home. What's to stop you from getting caught up in the system's clutter? What makes you think you have no chance? And so there are thousands of examples and questions with the same answer: "It is your thought and your attitude that brings you to what you make of life". Not everyone sees all opportunities, off course, but then those opportunities were not meant for him! So why does one person go through an alley and the other walks around? Listening to your feelings and unconditionally following what it says.

Fear, one of the biggest players in today's world. Everyone is afraid of a still unseen and undetermined virus; they are afraid of a ghost. People let themselves be lied to and injected with a substance that only Nano technicians know what it will do to your life! They take this product because they are afraid of a ghost and this manipulation is supposed to protect them! People truly believe in a placebo to save the world!

Well, if you look at the current times of fear and virus madness, then I indeed have a hard time seeing that the very thing that is being injected into many bodies is the killer, and that the so-called killer is nothing more than a flu that has been around for hundreds of years, and has now been given the name Covid. A positive Covid person, like a flu patient, has the same keystones but in the wrong place. That makes Covid very easy to cure! The great thing is that this supposedly deadly virus is in one place and stays there because it has no meaning in a human life. By activating it through the energy, it can start acting up, but it can never be lethal. 90% of the deaths are due to miscalculation of the regular doctors!

So much for this present time of our lives. Definitely not the easiest time, since many people collapse purely because of fear.

Chapter 4

OWN WORLD, OWN EXPERIENCE

An excerpt from my diary

4-1
Journal from the other world

It was October 28, 2020 when the message came after many months of uncertainty and searching. It was my pain management doctor helping me to make life a little more bearable.

I walked in and there were 5 people around me in a short time and I could not answer any of the questions. Before I realised it, I was lying on a bed with an infusion. The Doctor came to check on me regularly and when I was approachable, he told me he wanted me to have a brain scan. I refused, simply because I didn't want to confirm what I had known for a long time.

After the infusion I went home and went through a strange period. For the first time in my life I entered a world that was absolutely black. A black world without a single speck, not even a glimmer from my own eyes. I had apparently entered the world of absolute black! The moment I opened my eyes, they were hurting, I closed them and again I found myself in that world. The world of absolute black did not let go of anything, it was a world of absolute nothingness. It oppressed me, I called my wife and I cried out. I have never experienced this before.

As the day went on, the pains returned and the black world kept following me, even after I perceived the present world through my eyes again but did not comprehend it.

Once again lying on my bed, there was a moment where my thoughts went to the whole incident and the guides I have.

These guides are hanging as drawings in our bedroom and so "Kid", "Jeoreed", "Maunck" and "Lynquin" are with me every day. The latter I have not been able to contact for many years because apparently he was the last one in line. He obviously was the last guide who would guide me through the end of my life.

"Kid" taught me the basics of the Energenius life.

"Jeoreed" who explained earthly matters in this new world of energy.

"Maunck" was my mentor until recently and he was the guide that made me decide and allowed me to see the total life.

"Lynquin" was silent and unavailable until recently.

4-2
Black day

And then the day came when I saw absolute blackness and it was clear that "Maunck" was leaving me in this world without any guidance or explanation. Even though I had wanted to claim "Lynquin" before, it was now that his name popped up in my black world and I asked him what was happening here. Why am I being placed in a nothingness and why has my body given up on me?

And then Lynquin came with his explanation.

Firstly, he reassured me that I was not alone and that he was with me in this black world. Out of nowhere, there was a sparkle of light in my right eye, which exploded and I could see all kinds of blue dots. He said to me "that's me and I'm here if you need me and you know you're not alone". He then continued his story and started with quite a long story:

"John, from now on I will be the guide that will lead you through the final stage of a great journey. There is nothing more that I will teach you as you already know and have achieved all that you need on this final path. Other than that, do not worry about your body as it will not be needed soon and you will proceed in the energy and path you have chosen."

"You will have to deal with the very greatest in this world of Energy, since you are one of the few who have mastered this matter. Although there are still many paths to go, you will have to face the last steps of the resistance that is there now. The physician who brings you into this stage is going to help you even though he is one of the highest grades in the Masonic. Remember, you are 66 years old, but it is not for nothing that you have finished the 99 grade and are now moving on to the last 3 steps! This group will do everything to take away your knowledge, but most of all your energy for their personal use. You have to follow these steps to understand all about their powers, but also to stop this group permanently, and yes you have that power and you are one of few who will accomplish that."

"The black world is the world of fear, just as you lay there crying alongside your wife, not knowing what is going on. That is the power they work with and think they completely control. You will find out that what we have taught you, and also provided you with many proofs, is that fear cannot be overcome and does not actually exist. When Maunck withdrew and I did not respond, you felt abandoned. You started having doubts and you thought you were being swallowed up by the cult / devil. But still you recovered and you got in touch with me, and the path we are now walking, is partly in this body but mostly as energy. There are pains around this immense world of Energy. A lot of pains caused by wrong energies, but those pains are part of the source that everything is about."

"It was striking that an incident in an interview drew your attention. You immediately saw what we wanted to make you experience beyond life and what energy can do. When you stepped out of your body to show this man that distance is nothing, but also that you can be together 10 thousand kilometres away, you noticed that your body language and your eyes changed! You could clearly see the shifting and you were worried that it was a negative phenomenon, as described in the earthly scriptures. However, you witnessed your own ability and your true energy became visible. The human mind puts it under the standard reptillian and makes it a negative symbol, because it can indeed only be done by the great ones. You should know that it is there and what it is, if you want to continue in your goal you have set and we are here for you."

I interrupted Lynquin and asked who he really was because I knew who Kid, Jeoreed and Maunk were and did in my life, but in all those years I had a fourth guest who was silent and didn't say a thing. His answer was short: " Lynquin, are you John! The John in the world of Energy who has reached the 103 steps, who is above the earthly, has found the contact between the world of Energy and the world on the globe, and now being able to carry out things you have described for so long in your books!"

A very strange feeling came over me as I lay on my bed, while keeping my eyes closed, a pitch-black demonic world with a few blue specks had to tell me that I had overcome the earthly and have a connection with the world of Energy. It was strange, because it felt like I was being privileged. A feeling that is common among Masons who consider themselves elevated towards the divine. Written material shows that the "all-mighty" are the commanders of earthly life. They are between 100 and 103 grades. Lynquin talks about beings even above that, and what can be experienced and seen in the world of Energy.

Lynquin, however, continued and the next surprise came:
"John, you are this force that can stop the absolute top because you have understood that murders, wars and violence cannot disturb this absolute top. So Lynquin proceeded with telling me the way I should follow that does lead to success! It became clear to me that these absolute masters have, besides many powers, also their shortcomings, which explains why they are now pulling at me! According to Lynquin, they know that I possess the still unknown weapon, which cannot be answered by them. " From the

billions of writings in the past, there is a constant reference to it and also a constant search for it. Just look at the journeys into the worlds of the many religions, how people are diligently trying to get to Parallel Universes. Searching the formulas and the time gates, thinking to find the answer on the other side. Despite thousands of years of rooting, the ultimate answer has still not been found, nor will it be found, as you John has the answer and already applying it without knowing it."

Then I stopped Lynquin because it was beyond me and so I opened my eyes and noticed that it was already evening but he continued that I had started the new final path. This is the path I need to take in order to achieve things that seem impossible for people!

In rejecting any further information, Lynquin nevertheless rushed on and told me to draw a picture of three people, and that I should record every detail in a journal, as if I were travelling from another world.

To be honest, as I type this text I really feel like a fool who is recording a lot of nonsense. But I have been given the task of recording what I get through, which is exactly what I am doing in this book. One thing that is clear is that in the course of time, things will be clarified, something I experienced with previous books as well, because then I have already gone through certain things and situations and thus get explanations.
There is much that I myself do not understand and which Lynquin has warned me about. I was warned that this published book will only be understood in 100 years' time, and that it will still contain many mysteries.

4-3
More conversations with Lynquin

The conversations with Lynquin became more intense but also more detailed, in order to determine the path to follow. I first wanted to have a better body with less pain and with which it was easier to work. This was granted last night, and for a moment I found myself in the time of my first guide "Kid", who told me that it was possible to make contact again in order to repair the damage. It was like going back in time to recall some things. It is odd that after so many years, there are still things that slip your mind, and then leave you with doubts.

After the flashback, another normal night ensued and most of the pains were gone. That morning it seemed like I was back in the fight that needs to be fought.

That is where Lynquin came in again, telling me that all strength is needed now. It is not about the Covid hype but more about what is really going on, because humanity is going to make a 180 degree turn into a new era. Humankind as it is today is out of control, as it is clear that, again, people cannot manage freedom. It is assumed that freedom means that there are no paths to follow generally but everyone goes his or her own way, which is then set against each other. Freedom is a collective and not an individual path!

The fact that this has now emerged in 2020/2021 is because previous attempts failed. The reason is that the wrong strings of human life were being played. Now the right fear string has been struck and everyone has gone into a hypnotic state, unable to control and navigate themselves. One is dependent on the whole, which very cleverly makes advantage of this.

They have perfectly merged fear and hypnosis through certain frequencies, which many consider to be the 5G frequency! This is now coming out more often, especially by doctors and scientists who are involved in that field and who are slowly but surely being silenced! However, the evidence is there and from what I understood from Lynquin, I myself was under this spell for a while and hence my fears and disappointments in the past.

And now that I am back, I not only understand the picture but also the way it works. So now the important thing is to first tackle that, for yes, there must be an intervention in humanity that has gone out of control. We will not tolerate a total seizure of power against mankind, for then man would be nothing more than a labour horse or an energy generator for the great forces.

And there is Lynquin again, explaining to me where to find the weak energy points of the whole, but also, what the current forces are running on. It then strikes me that, despite the negative frequencies, I was clearly not far from the core that should be addressed. Is it therefore that I need the present doctor somewhere high up on the Masonic ladder, in order to find those intersections in this world of Energy? Obviously, there is a source here that I should know. Just as the connection came instantly, this must be a path that needs to reveal itself.

It was a day of many inspirations, also later on with setting up the drawing. This assignment, too, will certainly lead me to yet another answer to one of my many questions.

We rumble on, day after day, and somehow it seems as if everything has to stop. The nights are long and the days with much rest and retreat from this world.

All the time my brain wants to do something, I'm working on the Xirtam of which I now finally know the definition:

X people
I nternational
R einstall
T imeline
A rtificial intelligence
M anagement

It describes exactly what is going on and where the programme is operating. I have noticed that many people are hesitant about Xirtam, despite the fact that we now know that the world is completely governed and controlled by Artificial Intelligence. Then when you call it Xirtam and refer to the incidents as such, people are incredulous and smile.

On the subject of non-belief, I had quite an outburst today. When it comes to publishing, people keep asking for proof and written assertions. It is odd that a statement by me is not seen as a witness statement! Giving "first-hand information" is something that one seems hard to swallow, and even though I then suggest that I state it under oath, to many it is still not a valid declaration.

That is strange, because when a witness sees a crime and appears in court, this statement is accepted and considered true. Whenever I am confronted with statements, and in this case my books, one says that there must be evidence. This is something that always happens when my name is involved. The constant denigration and calling it a false statement does make you tired at times. Apparently, not even a judicial declaration is enough then!

So this morning, that just blew up and I immediately went public throughout the social media that everything I post can be confirmed under oath.

When fatigue is a big issue, you also need to fight some of those mindless, non-thinking creatures. Just a waste of time and yet another annoyance. The funny thing is that, even though it is difficult, I pick up my daily life and try to share the most important information that still come in daily, via MKK or as private messages. Unfortunately, as has been the case all my life, I am not understood, and for the outside world there is only one goal, and that is to destroy me, or rather to eliminate me!

Chapter 5

The decomposition of time

5-1
The description of time

This description is not the one we know from a clock with 24 hours in a day, day and night, etc. etc.

Meanwhile, we know that time is a manipulated human constraint.

Then questions arise such as:
-People only live for a certain time and then there will be others who go on.
-There is a "past" and there is a "present", both of which are in many cases totally different.
-So what is it that people can travel through time and switch between the past and the present and the future?
-There is a definite day and night and they have been following each other endlessly since the earth existed.

Please let us dissect the time completely according to the questions/questions.

At first it seemed that there was still much to explain, but it became increasingly clear that we had really been fooled, and that the truth was still difficult to see.
Now then, there we go, dissecting the word time.

5-2
Question 1

-People live only for a certain time and then there will be others who go on.

Answer

We consider life to be a period of time in which we are allowed to be here on earth. For several people, after a number of years, it is the "end of life". In reality, it is not an end, but, as they sometimes say in faith, "passing over"! Passing over is returning to the true source which always exists. Then you might say "well, then you are talking about time" and I should say "no", for it is a limitless form of energy and not linked to any solid matter or timeframe. Hence, one person has 12 years to live and the other 100 years in earthly time measurement. Now then, 100 years in the world of energy is equal to those 12 years, since everything is in alignment. Not one line after the other as was always assumed, but a side-by-side line and this will determine the "time travel", which we are going to answer in the question 5-4.

There is a continuous accumulation of reactions in the World of Energy that keeps life going, but just as well ends the life of a 10-year old person. All in order to get that reaction that is needed in the whole! Consider every human being as a spark plug that provides the spark needed to make the machines move! Any particular part of life depends on the energy it receives. Unless you see life as simply "being present in a particular moment", you will not notice other doors opening. You are much more concerned with the earthly and not the whole. To detach from time is to "experience more" in your life and recognize things that others never experience. When you are influenced by time, your world is narrow and limited what your energy can and may allow.

5-3
Question 2

-There is a "past" and there is a "present", both of which are in many cases totally different.

Answer

The difference between the past and the present is, our "seeing" in the World of Energy. For instance, many people do not see those buildings in the clouds at sea, simply because they do not have that ability. So, there are what we call "witches" who see more and know more than the masses. And then there are those so-called "scholars" who think they see and know more through explanations. However, in many cases these are dead-end channels that limit themselves by basing their arguments on number and time.

Furthermore, those people who are highly gifted are important, because they often unwittingly get their information from other sectors from the World of Energy or as we sometimes say "from other dimensions". That is why these persons achieve amazing things, because they receive information from "other worlds", as they tend to call it. As a matter of fact, they have the ability to dig deeper into the energies and thus enter into other dimensions. The knowledge they gain there, they take back into the earthly energy and that is why these people have a supernatural ability.

Unfortunately, many gifted people are taken in by multinationals and corporations where their talents can do nothing more than contribute to some flow of money, power or domination. Only a few gifted people have the chance to develop naturally and achieve extraterrestrial achievements. Controlling this group of people means that, as a current human race, we are going backwards rather than forwards in many scientific fields. Many are permitted to investigate only a limited number of cases and it is often completely covered by certain people. For example, all the books and know-how that are kept in the Vatican but cannot be read because it would damage the Church. Nowadays, this is starting to weaken slightly as more and more writings are released by people outside the church and thus various aspects of the energy are being accepted more and more.

Opening up this know-how also makes it easier for people to talk but also to write about what is really going on around them. The UFOs are no longer a fantasy and various theories can now be read and used to explain things that have been distorted by faith.

Which brings us to the explanation of the next question.

5-4
Question 3

-So what is it that people can travel through time and switch between the past and the present and the future?

Answer
Time travelling is actually shifting between various sources of energy that we have available around us. It is not necessarily a "gate" that you have to pass through. This can simply be a self-made entrance to "escape" from the block in which one has placed one's own life. By creating a portal or a transition, one is experiencing a past or a future because the the World of Energy cannot explain where one is. In fact, it is the same energy, the same One, and not a different time or place. Therefore, we see images from the past of spacecrafts but also of the portable telephone, etc. These are images from other blocks that sometimes merge with what we now experience and believe we see. According to the drawings, rockets were also available in the Stone Age and Hitler's soldiers had portable devices! This because there are mergers of views of the highly gifted who are able to see the world of the many time periods as being one energy, one time and one life.

I said "think to see" because every person sees the world differently and consequently experiences it in a different way. This cannot be otherwise, because everyone has his/her own energy block and is performing the work that needs to be done in the World of Energy. The human's simple sources of energy linked to the total World of Energy. So, if you travel from world A to B, in your eyes they are different worlds but actually it is all the now, aligned from a different personal energy source, a different perspective.

Not easy to understand, simply because the brain can't imagine it, not to mention recognise it. This is a totally different way of looking at things, but also a different way of thinking, as we can see in Quantum Science and its energy.

5-5
Question 4

-There is a definite day and night and they have been following each other endlessly since the earth existed.

Answer
The day and night are changeable, as is the timekeeping, which has to be adjusted every year, otherwise it will go wrong according to the timekeepers. The endlessly repeating day and night are not due to time but to the earth's rotation around the sun and its own axis. It is quite simple, when the earth comes to a standstill, it is end of story for a day and a night. You can see this at the North Pole, where there is no light for many days and where the night persists. The end of day and night measurements. So they have clearly taken the day and night standard to introduce time.

Nevertheless, time is still linked to this, but since the last few centuries, the rotation of the earth has also been involved! In other words, it is ratty and they really do not know how to cover up the lie. Since all science to this day calculates and bases everything on time and distance, no one would dream of deleting the time/distance aspect. As I stated earlier; Yes, there are changes and especially in Quantum Science there are big changes and it turns out that time and distance do not exist or can not really be explained.

The Quantum is aware that there is a great unknown force that is thought to be found in the "dark matter", but also that time measurements are no longer correct as many measurements are the same at a point A and B, even though they are thousands of kilometres apart.

Therefore, you can see that distance and time are not important factors in true science.

5-6
Art

I just saw an artwork by a person who paints illusions on large walls. Now you might wonder what that has to do with all of this? But to see that such a person can take you, on a flat surface into an abyss, another building, or another world, then you see what it can do to one's brain and one's perception of things. In this way I would like to show that actually everyone can see things that are different if you look at it from a different angle. These works of art are a good example, because in one place you see a completely different world. When you reposition yourself, it is a flat image and a distorted picture!

However, in the world of energy, the images remain three-dimensional and so those who have the ability see it as a full image.

5-7
Spirits, Ghosts and Phenomena

Then there is the world of "spirits" and "ghosts", which represent energies that are lost or that, in other forms, have to stay for many earthly years to complete things. Such energies are in abundance because many of them do not achieve their goals in their world of energy. They wander and seek for some attention in order to complete their path. Several worlds can meet in those ranks, and chairs or cabinet doors can easily be moved and seen in both worlds! But one can continue in this way by manipulating but also adjusting things. And there you have the conflict between dimensions, which can sometimes seem awkward and difficult.

Since this branch of energy is very fearful and not understood, it is mostly placed under witchcraft and demons. It is a pity that a religion propagates this as being the devil and evil spirits. There is none of that if you can understand the world of energy.

The wonderful thing about the World of Energy is, that by controlling and seeing the energy, the given "no time, no distance" and thus this "narrow" branch of life can be explained. You then realise that it is possible to work with it. There is much to clarify as soon as you adjust your mind and allow what is really going on around you to come in.

5-8
Today's world and the 180 degrees turnaround

The year 2021, more than one and a half years after the fake virus Covid, simple flu virus, was launched, humanity has reached the point where there are two parties in this world. The frightened people who think they are going to die from this non-existent virus and the ones who don't participate in this nonsense and realise that it is a political move to change humanity.

People think that vaccination will get rid of this virus, but the vaccinated people actually die from the nanotechnology that is injected into them. I will not go too deep into this matter as I have already written at length about it in my book "Project Covid", but why bring it up is that we can clearly see during this time that people are very easy to manipulate and can even be made to believe that they will die if they have not had their shot!

This example of Covid is what has been going on with humanity for thousands of years. People believe stories of others and are unable to rely on anything they see, hear or feel for themselves! They dismiss everything that is not presented to them, as nonsense or lies, as we now see in a social media where everything is directed through a so-called "Fact Checker"! People assume this to be the truth, not realising that this is the very same party that wants to reduce humanity to half a billion! Now we see that under a Covid label millions of people are dying in India, while in reality it is from starvation because all supplies of food have been blocked in order to eliminate the people there. It also turned out later that the animals in those places died massively, and this is clearly the result of the 5G being activated in those places!

The point here is that in many cases people blindly assume things because they have lost the ability to see and think for themselves or are not taught to do so. In education, we also hear that children are being taught less and less. All in order to make the children think less and be less able to see for themselves.

In this world where time determines everything, it is now clear that time was already a precursor to this manipulation and that the next step is a clear demolition of human existence in its old form. For if man no longer has the ability to see beyond the standard, these creatures have no choice but to act as robots whose energy is taken, not for general energenius use, but for the manipulation of mankind.

As I see it today, the brainwashing of mankind is truly beyond words. But even worse, is this hypnosis that drives people against one another. No more is war a war of tanks and shooting. Instead, war is people killing each other in their daily lives. There are large groups of middle-class people that are disappearing, while other groups are becoming huge, controlled by robots that have been completely brainwashed and are happy to have a slave job in that structure.

5-9
Frequencies

5-9a
5G

The most serious is the emergence of the 5G that has been placed worldwide and that determines the life and energy of mankind today through frequencies. While it was Hitler who experimented with it during his time, it is now a worldwide phenomenon that causes many people in unexplained ways to die. As I mentioned, in India but also Saudi Arabia, Wuhan-China and even Italy and several other parts of the world people are falling down dead by the dozen. It is striking that this is happening everywhere where 5G has been installed and made operational with utmost speed!

And this brings us into time, into the part of frequency manipulation. Because frequencies are an important factor in the present time measurements and they therefore indirectly and directly influence our energy manipulation c.q. time perception!

So I would like to make a personal point again:

5-9b
Those little signs

At some point, I fell into a negative spiral when I constantly encountered those people and their problems. For 66 years I had managed to protect myself against this negative energy, then suddenly this changed and I found myself in the path of death and pain. When I asked my guide Lynquin for help, he did not get much further than the words "you have to experience it" and "you are going to have great times".

It was like being betrayed, fooled, but mostly it was the "experiment" that hurt the most.

The experiment

How is it possible that mankind is so numb and hypnotised?
Having themselves injected with an experimental substance?
How is it possible you let your own life be taken away by and through an experiment of the system?
It is known that about 6.5 billion people must die!
Just how can one give permission to experiment on one's own body?
But how is it possible that people think they can be free when they are injected with technology that enables a system to decide whether you can have a child or not, have a job or live on the streets? After all the bad education and deliberately keeping people stupid, are there so few people left who can think?
But what is it that prevents people's eyes from opening and makes them lined up like slaves waiting for their fate, their death?

There are hundreds of questions that simply prove that there are many bodies already owned by the system and acting as robots promising to travel and go to pubs!

And a body that has been put under hypnosis and the system promises you freedom and life!

You already died like sheep on a slaughter bench.

It made me sad when I published it, but more than that, it was the message it carried to the whole world. That impulse, that one spark was for me, that the world also saw what is actually happening around me. Helpless and observing a world problem, I suddenly became aware that a large part of the world's population would be lost through the injection, not of an experimental substance (one knows what the substance is supposed to do), but a substance of which one was certain that then that person would be controllable for the rest of his life. The intention was to create two parties in order to pull the human strengths apart. The absorption of all human energies had been set in motion and this made it appear that it was a kind of mass hypnosis, which in reality was the draining out of the last remnants of energy of each injected person. At the moment, the human robot is on the verge of its first birth. The controlling authority is not going to eliminate everyone right away, but first collect the last bits of energy. The rights of having children, the rights of freedoms have been taken away. However, the rights of life depend on what you, as an individual, can contribute to the whole!

This may sound like science fiction, but it has become reality, and was confirmed by the break-up of the world's richest couple, Melinda and Bill Gates. It is obvious that there is a lot of friction between the two of them. And all signs are that Linda clearly had a problem of conscience when she found out that mankind is going to be completely controlled by Bill, all through his invented technique.

5-9c
Loop

N now you may ask again, what does that have to do with this book? And my answer would be ''everything''! Throughout the many thousands of years of earth's history, we see that, in fact, everything is a repetition, and that, as a matter of fact, nothing changes. Now you might think that we as humans are in a "loop" which is very close, but it could also mean, in the theory of no "time", that we humans may only have limited information. It seems that we have to complete the same programme over and over again in different guises. When you translate this into the world of dimensions, as well as into the world of a hologram, it then becomes more clear that we have nothing but toys and have to create our world with them and manage to survive. As a hamster in its cage in its running wheel, watches endlessly a part of the world. It has to maintain its energy for the one who put it in that cage. There are also people who do not go further than a few metres from their home, people who live in the same house all their life and others who take a country as their beacon or wander around the world, which they call "adventuring". It is determined per person, per case, where your energy is needed and there you will stay until the body no longer generates energy.

The 'settling down' effect is programmed into every human being and as soon as you deviate from that, problems start to appear in the whole programme. Because time does not exist, we live in a loop and we are nothing more than a Duracell battery that provides electricity for a certain period.

Is that life?

Broadly speaking, yes, because many people on this earth do nothing with their lives. Worrying about money, house and all that comes with it, and others fighting for a living, food and when possible a lot of fun. And yet that does not happen and many people pretend how lucky they are and have had an amazing life. These are dreams and concoctions that we have all been given because we should not get too deep into the energy dip. People who have major depressions and who think differently, as well as artists and those who see more, are not fitting into the times. They are outsiders, difficult and unwilling to share their energy with the existing system.

Then there are the so-called hyper people, overactive and thinking that there are 3 batteries in one body. What they call ADHD is a distortion, because there is something else going on in those bodies. They want to do things, to perform and have a pool of energy. Most of the time, it is the substance sugar that makes the body overactive. If you exclude sugar, the locomotive comes to a standstill.

We can see time as an indication that has been dictated by mankind in order to get people in line. In the beginning it was fear and suppression of the people, which later was seen that the people were not doing their part for the benefit of the higher. 'Time' was one of the first tools to force people to stick to a so-called fixed rule! With the concept of time came watches, clocks and timekeepers in order to punish people when they were not on time! After that many other devices were introduced and one of the most well-known in our present life is the computer. A device that can do everything for you! With the current supercomputers that are linked to the new way of quantum, they really think they have found a replacement for humans.

And then there was the catch, humans are no longer there to do the work and combine forces to generate energy. Instead, man became useless, an-noying, thinking too much and eating up all the resources like an animal! Now we are in 2021 and man is being disposed of ingeniously, thinking that they can live longer and are the chosen ones!

The chosen ones have disappeared into the new world long ago and have established their existence there for many years. All the troublesome and energy-consuming creatures must be restrained. However, China and India are handling it more resolutely and are murdering entire states by not supplying food, a process that has been going on for decades in Africa and South America! Humans have become a plague of rats and all you can do is eradicate them!

Of course, it is not the most beautiful message and not the most inspiring text, but we need to face up to that as well. I simply wanted to highlight the fact that it is important not to play the system's game. Vaccinating is for the stupid ones who want to kill themselves! Showing up at times set by a system, is what you don't respond to. And by doing so, you are going to create another world for yourself, where a system has no control! When you keep your energy pure and as little as possible polluted with systematic weapons of destruction, you will re-generate the pure energy to which the system has no answer.

Just an off-the-grid example:
Have you ever watched a superhero cartoon or film? The bad and the good creatures. They constantly show flashes of light, supernatural powers and the clustering of forces. In reality, the story is about showing people that the bad group wants to take away the energy of the good group in order to dominate the world! It has been told hundreds of thousands of times and still almost nobody understands it, apparently!

By distancing ourselves from the current system (bad movie characters), we as pure people can gain energy to which the bad group has no answer! Moreover, for thousands of years, the groups that now work under Freemasonry have been trying to possess and manage these powers through ritual. With all their books and know-how, however, they have not really made any progress and still do not know how to handle the true powers, not to mention possess them. These true powers should not be misused!

In other words, see your life as a cartoon or a hero film, with positive input, independent of a corrupt system and not working in terms of time, you can gather all forces for the good.

Remember your heroic movie! It is a true fact!

5-10
Energy, impulse and frequency in a block called 'time'!

In the previous section I wrote about "absorbing human energies".

For years they have been trying to steal energy from people and so influence their behaviour and their lives. Unfortunately, they do not realise that time is linked to frequencies but also to energy. So it seems that since ancient times people have tried to influence the world of energy that surrounds humans by using 'time'. The impact was nihil despite the fact that there has been a turnaround since introducing time into human life.

We have now entered the quantum period and it is only now that we can clearly see what time is, or rather, is not. With this in mind, China has been working very hard in recent years to control mankind and consequently to control the whole. It became obvious that computers were needed for this, because with enforcement they couldn't achieve total control and the first quantum computer was developed, which had to map all commands but also all energy sources. Now that China is largely under total control, it is time for the rest of the world to join in, although there is some resistance, but with the introduction of frequency manipulation called G5, they are able to make a breakthrough into this domination.

Time is an important player in this manipulation and despite the fact that Quantum sees 'time' differently and it is not a fixed fact, they have managed to link this manipulation to time as well. Thanks to this method of registering and processing everything, a mega leak has appeared in the entire world domination which is 'time'. 'Time' that has been set as a standard for the current systems, is therefore very easy to break through and circumvent, when one is aware that 'time' does not exist and can be influenced by energy and frequencies!

This is where the true link can be found and used to make computers of billions, but also the newest techniques, go completely off the rails when one continues to use the concept of 'time'! In the world of energy everything is possible since we are not linked to anything, and there is no front, back, left and right, nor up and down! Everything is around you and is one with the space you are in and can be manipulated. This I have tested several times and it works! Both mega companies and large national organisations will go into error if you mess in their core information with the concept of 'time' through frequency.

Haven't people been afraid of the millennium problem? The Bios could go crazy and computers unusable! Have they gone off the rails? We are told they did not, but is that true? After the year 2000, what have we seen and what is going on worldwide? Many things have happened. And the world clock had to be fixed again because it was not running on time!

Can you name one natural phenomenon that does not work flawlessly? That would explain, and also prove, that time is handmade and not a product of Mother Nature!

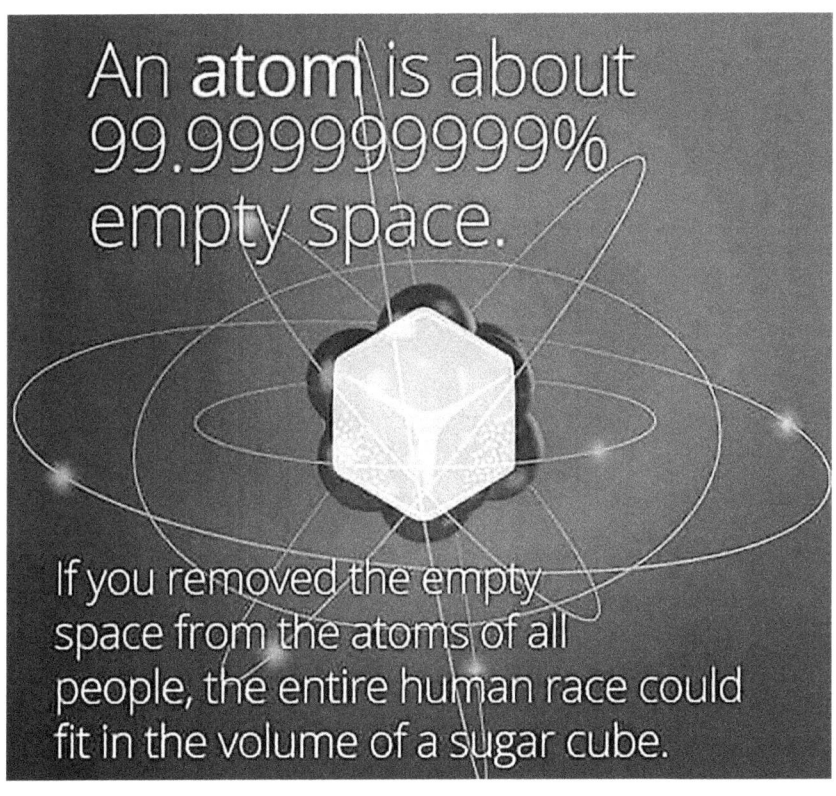

An **atom** is about 99.9999999999% empty space.

If you removed the empty space from the atoms of all people, the entire human race could fit in the volume of a sugar cube.

Chapter 6

Now - Illusion

6-1
Time, the process of manipulation

Many pages have now been written about time. Reading the text afterwards, I personally think it has been given too much attention! But that's the way it is, and I don't tamper with texts just because it could be better. It is just as it is written and according to what I have been told. Although I have my doubts myself sometimes, because it does not fit in this current time. Then I think of Lynquin's statement, who made it clear to me that this book is yet to be written, but will also be a mystery to many people until the proper time is there.

But let us recapitulate for a moment and clarify some things:
- The theorem "time and the distance connected to it do not exist".
- It has been described in great detail and in today's quantum world it is being proven time after time.
- Should time and distance be a product of nature, like the entire Universe is built, then time would not be "adjusted" yearly. There would also be no strange formulas that are constantly being added to calculations in order to give time the prominent role. Also, when calculating distances, time and again it is a factor, that has to be added via obscure tables, in order to supposedly determine the true distance.
- There is absolutely no knowledge of the true distances between the planets and stars, and to calculate the universe with its many systems is to guess.

- Scientists questioned their own calculations after a while, as they realised that the basic time and distance was wrong and remained a dark area to this day.
- The most important question mark is the proof that the "nothingness" is "something" and that energy is many times stronger than the "something"! The by us labeled "nothingness" is an empty bucket filled with the unknown!
- However, one sees one's world this way and the other sees a completely different scenario of a life!
- Many people are aware of an energy but interpret it as a nothing that in spite of that the energy, the "nothingness", makes the whole.

That makes it impossible for you to get out of your block (life), which is called time in earthly terms. However, actually you live in a complete world of energy which is the engine in all places so that it cannot be destroyed or abused! The ingenuity of this world of energy is that it is self-healing (autonomous) and even a rotten apple like earthly mankind has no grip on it. By isolating it, it remains stuck in the whole and is insignificant and irrelevant.

See it as your own body which is its own Universe and can therefore easily eliminate diseases through an immune system! We therefore call it an autonomous body. In the current earthly time, people think that by messing around in an immune system, they can dominate and manipulate people. This battle has just begun and there are already "anti-channels/energy" found in those people who can work with the universal energy! Indeed, the manipulation in a DNA - RNA and your immune system, is a war that the body, fed with the pure energy c.q. frequencies, will win. The evidences have started to come from certain angles.

Many questions are then answered and various worlds can be explained.

6-2
Let me just mention something that is going on right now

Nowadays, we see that many very famous and high-ranking people "disappear" but also supposedly die, and many more work with doubles. There is a lot of writing about this and many pictures and examples are posted. In the book "Het lang verzwegen geheim" I have explained everything about a "second world" as well as the UFOs and the various groups that are way ahead of us. For more than a hundred years, the "new world" has been in operation and is gaining ground for those people who have only now realised that they are alive and not alone. This is one of the last stages of what we call "shifting" and which is going on now. People are disappearing, those people who know too much about this earth or think they are still "useful" for the whole. Not only scientists but also great and famous people are being separated from this world.

The clubs they have been in became a threat to humanity, and we as ordinary "batteries" are not permitted to see this know-how in its entirety. Think of all the people who are now supposedly being arrested for child abuse, the wars between the churches, and the Vatican that remains silent and which is now joining forces with all the other religions! See the so-called execution of these great men of the world who worked by the rules of their club. See the so-called execution of these great men of the world who worked by the rules of their club. All have passed from the earthly to a greater life to protect the source they know.

All of these things have to do with the fact that now too much has been revealed that was kept secret for centuries. Think of the many papers of secret services, the disclosure of UFO's and the admittance of many actions against humanity. Everything has been released in a very short time and many prominent people with a double have been convicted, murdered or disappeared!

The world today has been released to those who are strongest. It is China that has taken control of all the world's economies, as well as financially controlling almost every country. A mega debt is anchored in that country! Therefore, there must be a shadow economy and residence, as described in the book just mentioned.

Time is what we were talking about, and in the current situation, too, you see that time is playing a double role. On the one hand, people talk about time being too short; on the other hand, people say that everything will change in 2025. Well, both are just talking as the changes are happening in full swing and because of the mass fear and hypnosis and the big 5G frequency pollution people are not aware of the situation they are in. The complete blindness started years ago and through food, frequencies and through spreading fear, there is now a world population that does not see that a totally new dimension has arisen where mankind no longer plays a role as they used to. The humankind is stuck on lies and blinded by falsehood which the brain is unable to dictate as being "not true".

I would like to talk about the little group of people who call themselves lightworkers and who make me shiver when I see how they think they can save the world by performing "na na na"

It is important to know which frequencies to use and a few tones in certain frequencies and some light is not going to bring the earth into the higher realms. People need to wake up and you don't do that by bringing them further into sleep/trans. That is only for a few and not the way to awakening. The important thing is to connect with the true energy, where the many clubs also get their power from, but thankfully do not go beyond the basis. This is what I call 'thankfully', because if these clubs were to find the real power and start working in the world of energy, then we would be lost as humanity. However, fortunately that cannot happen, because the safety mechanism (Autonomous Machine) has been built very solidly. There is nothing a human hand can do to interfere! There is an opposite for every know-how one believes he has, which can be found and worked with, moulded and manipulated.

There is Yin / Yang, black / white and a world filled with many places that open up as soon as you know how to use the right combination.

6-3
Seeing or seeing through?

Over and over again I write about seeing the energy and working with it. But might it not be that in everyday life, it is seeing through the other person that you are facing?

When looking at the energy, it is rather that you do not see everything directly as a chair, table or any other object, but instead you see the building blocks of everything around you. Meaning also that you do not see animals and people or any other living creature, as well as trees, plants, sea and their lives there as one single object. There again, you can see the whole structure (building blocks) with all their structures and on which they live and are built. Because you can see the energy, you know where you can knead and adjust, and also where the disturbing elements are.

Therefore, seeing through things has indirectly to do with seeing the elements. However, most of the time, seeing through situations as well as the person in front of you, has to do with feeling. Since time does not exist, that feeling is something one works with and makes decisions whether to accept certain things and reject others. Understanding situations is more about working through a subconscious mind that gets its impulses from seeing the true energy.

6-4
Time, an illusion of mankind

Time is an illusion of mankind which imposes many constraints on it.
- One needs to be there at that time.
- You have to eat at this hour.
- You need to do this and you need to do that and all according to a time schedule.

Likewise, we know that we have to drive 12 kilometres to reach a destination, and meanwhile the moon is 384,400 km and the sun 151.56 million kilometres away! Mind you, all these numbers are obtained by formulas or calculations using a certain table.

Through quantum determinations we know that the true numbers can sometimes turn out strange or completely different and we know that by using the old calculations it is still impossible to go faster than the speed of light (1 billion kilometres per hour or 300,000 kilometres in 1 second)! This is another one of those limitations that has been forced upon us for many years. Incredible numbers that undermine all logic and all reality.

In this part I stated in the heading of the chapter that time/distance is an illusion and this is something we see being confirmed in our quantum and CERN experiments. It is not light, according to many quantum theories, but time that turns out to work much more closely with illusion since the same stimuli as well as the same data are being observed at the same time and there are no time differences anymore. These measurements no longer matter, regardless of where on earth they are.

This brings us back to the "nothingness" which is clearly something, and therefore a great conductor of all that is time and distance. Since we link the "nothingness" to nothing, distances are ridiculous and galaxies are mega light years away and the universe is 93 billion light years in diameter! Is that credible? No, simply because this number is made up! All these extremely high numbers are nothing more than scientific aberrations. These numbers are wrong because one categorises the "nothingness" itself as nothing! One is even sure of an age of 17.7 billion years and that while one still sees the largest matter as a "nothingness"!

There is always the factor "ego, power and possession" revolving around science as data, we as humans place all formulas around ourselves! Just place humanity completely out of everything and place Nature at the centre of everything, and you will see that time and distance have really no basis. Mother Nature does not work with time, but with the element darkness and light, strength and weakness, here and elsewhere!

That from here and elsewhere you should see over the whole of the Universe, for the tree at the back of the garden has connection with everything, at the same point of life of the whole. To the energy, which is a force, it does not matter where it is.

The illusion is what we live in.
- The illusion of us humans, the power, distance and limited brainpower.
- The illusion of living life as we see it.
- The illusion of thinking we can control the overall forces.
- The illusion of the things we see!
The 99.9% of what we see is the world as we like it to be, it is just that 0.1% that we feel but do not see. Are you seeing and feeling more than that, then you belong to the group of "madmen and witches" and on the dangerous side of life! However, seeing things can provide answers. Indeed, some scientists are guided by "inspirations" but also by formulas that then arise and are hard to explain. This is how I see my way of writing down things of which I do not have all the knowledge myself, but which are dictated to me from the world outside.

In this present world, having seen the whole, it has become clear to me that I do not want to be here any longer. The worldwide experiment on humanity is against all energy. You don't do experiments for then you know it is a manipulated science that is not based on advancing humanity, but is based on the principles of a group that is driven by ego, power and possession. Therefore, it is important to stay far away from this and not allow your energy to be abused by any power.

6-5
Should time be seen in a different form?

Is it possible that man interprets "time" all wrong?

Time was invented by man purely to get people in line, that is quite obvious. If "time" were something natural, there would not be so many gaps in the determination of time and people would not use such ridiculous numbers when talking about distance. Everyone talks about "living in the now" and if the "now" is one line, then it means that everything can be manipulated! This also shows us what energy does and together with this, we come closer to the theory that the "nothingness" is much more compact than the something! This explains earthly time travel, as well as the going back and forth of several people in various worlds! Things seem to be easier to understand if we see "time" differently. If we start seeing "time" as energy, then wouldn't we be on a better path and wouldn't we get more answers?

We have witnessed so much already in the earthly time regulations that it ought to be clear, that "time" is not a natural phenomenon. When we start seeing "time" in a different context, we realise more and more that many worlds are very close to one another.

- The city that I see in the sea, is what I see now.
- People being seen and then disappearing immediately is not hocus pocus!
- The fact that people see UFOs is not a fairy tale.
- The events that keep on repeating and are seen under "déjà vu" are not a coincidence.
- And even the karma in many beliefs is not a fantasy.
- Neither is the energy I see a dream.

There are many examples of playing with time, distance and what one thinks must be past, present and future. Do not forget, our brain cannot process much because it does not know the whole information. How are you going to name something if you don't know what it is and what it does in the big picture? And this brings us back to the point where mankind is deliberately kept dumb and is given a little bit of information each time. The moment to completely change science has not yet been achieved, despite the fact that we have had the tools in front of us for quite some time. But despite that, we have once again plunged into a direction of computers, bits and an internet that gives a completely confused picture. Today's electronic technology is completely out of control and people no longer know how to separate many things from one another, and they create their own worlds as well as their own laws and rules. Rules because it is mentioned there or there and it is no longer possible to find out where the wisdom comes from. Those who no longer think for themselves but blindly accept electronic information as true have now taken over. This is a dangerous situation that we are seeing now, where people blindly assume that vaccines must be administered and where a supposedly dangerous virus is also haunting the area. The age of 2020 has shown that around 80% of the population allow themselves to be chewed up by everything, without thinking or looking at it from various angles. This culture of fear is so enormous that, for the most part, we as humanity will be falling back very far. Exactly as planned, where human robots give their lives to an artificial world.

This whole upheaval is due to the fact that people have let themselves be manipulated by time and distance! For many years, people have gradually broken down their own thinking and understanding. The imposing of many rules, as well as the taking away of time due to working hard in order to have enough food and a roof, has been creeping in over the years. We clearly have the 'settling down' culture and one has to travel and must be free! The latter has gone so far that the freedom people talk about is actually a prison for them. There is hardly any time left to do anything oneself and our lives are determined by a system that has been meticulously designed so that everyone is dependent on it. Just look at the system of technology where many people are addicted to social media, games, gambling and living in a virtual world! Have your friends leave their portable in the hall and try talking to them for 2-3 hours! You will hardly succeed, as they are addicted and linked to the artificial world that owns the whole world.

Day and night they can follow you through these smartphones, simply because they are on the internet both day and night! Other than that, it is their Bluetooth and also their 5G that allow the system to track how you live, work and what you do with your free time! Not to mention everything you discuss but also request and receive and put in your profile! Just type into your phone "Hassan, how do I get an atom bomb?" and the bells and whistles in the system will start ringing! The control is enormous, purely because a lot is already processed in a quantum system and the 'possibilities' are calculated, not on time, distance but on the basis of the 'present'!

6-6
5G - Morgellons - Nano and AI

As we enter the world of AI (Artificial Intelligence), it is becoming very interesting. We now know that the latest quantum computers work in various dimensions and thus determine the path that this computer thinks it should take. So the self-thinking mechanism also called "Xirtam" is a machine that determines its own path.

To build a machine like that is something remarkable, since the quantum computer is the first computer of its kind. That is because it is the first machine that works with DNA! The quantum computer stores all its data in a 3 string DNA. Humans have a 2 string DNA and with this machine a third one has been added in order to play with the human RNA and DNA!

I can already sense it coming that people will declare me totally insane, but I write it down here anyway because it has to do with the world or worlds that have no time.

The latest quantum computers that have a storage in a 3 string DNA (used to be hard disk) are linked to human RNA and DNA and work as a kind of GNA or free software that can determine its own path! There is also a kind of "wayback machine" that can retrieve old data, as long as it is digital, and place it where it belongs. In this way, the quantum computer obtains a large and clear overview of everything that is human and events. This connection is made via the long-established Nano technology, which has been incorporated into food bit by bit for decades! In about 2000, the morgellons disease emerged and is related to unexplained (until recently) fibres and particles that go under the skin!

These morgollon fibres and particles have now become outdated, having been brought into the body through Nano technology and through food!

Those particles that react more strongly in some people than in others are needed to build the Artificial Intelligence era, just as in 1981 that IBM was commissioned to build the first quantum computer. It was finally in 2017 that the first computer was presented, and enough material had been put into people in order to start working on the computer that works with the DNA principle.

For now:
You might ask what does this quantum computer have to do with time and also with DNA and RNA?

The human body is an electronic source. For example, I personally cannot work with a mouse or keyboard via Bluetooth because these electronic devices, can't cope with my frequencies and electrical charges. For example, I personally cannot work with a mouse or keyboard via Bluetooth because these electronic devices, can't cope with my frequencies and electrical waves. I have had computers break down in strange ways and years ago I had a Dell server where Dell could not determine why the main chips had failed. The report said that strange frequencies were interfering with the chips!

Since man lives through frequencies and is an electronic body which is a perfect computer through water, the quantum computer has been developed in the same way. This quantum computer can be connected to the human being, which is actually what happens to every human being who has enough fibres and particles in his body via the Nano technology.

I have published it once and there was laughter.
I am posting it here:

Bluetooth, 5G and frequencies

It has been proven many times that with the initiation of the current global farce, it is about an RNA manipulation linked to an advanced yet experimental control of humanity.

It was China that first did these experiments on its people and the world was led to believe that it was about face recognition and monitoring criminals. Rapidly, the whole of China was under control and they realised that cameras were no longer sufficient! It was necessary to find a way to track people day and night wherever they are!

Thus, nanotechnology was added and around the years 2015 - 2018, people were injected en masse with the first RNA Nano Recognition Chip via flu shots as well as vaccinations. Since then, people have been observed day and night through this technique. The cameras gradually became superfluous, but they could not be disconnected because they did not want to reveal this particular technology right away, so they supposedly continued to perfect it.

In the meantime, experiments were done with bluetooth, which was not a success as many people did not have a smartphone! The experiments were extended via 5 G networks. In that way, many people were eliminated! That was back in the days when people simulated a so-called virus outbreak in 2019. They were forced to hide their control weapon urgently and they did so by means of a new virus.

New telecommunications networks were installed worldwide on a large scale to enable nanotechnology to work there too. By launching a worldwide pandemic, they plan to control 80 to 85 % of the world's population day and night and, according to a Dutch minister, "slowly but surely to sicken and clean up the population". These experiments we now see in places where the population, through various injections as well as food, and having had the Nano control chip in their bodies for a long time, are slowly dying by the dozen!

World War III has been in the making for years and now it is being waged on a global scale.

John H Baselmans-Oracle

So much for the piece, it seems I was very close to the truth.

The whole theory of:
Quantum computers, Nano, DNA, RNA, 5G, Bluetooth explains exactly what people are working on around 2020!

"Breaking human DNA"

It appears that too little Nano material enters through food to drain people of their energy and abuse their bodies! The new generation of quantum computers needs much more energy/DNA to keep developing! Therefore, the world is rushing to connect people via a fake virus, in order to make the entire artificial system run optimally! That is why Bill Gates is in the health business and has access to human energy!

We may say that the big computer magnates use human energy for their computers! The statement Bill Gates as "health guru"! It is obvious that the elderly cannot generate much energy and are therefore superfluous, but meanwhile the youth, no matter how young, is the largest human frequency provider for the mega-systems! Could it be that the system we keep talking about, is actually a computer?

6-7
So, what is really the system?

Without knowing it, we have just travelled through time. You have seen what is really going on and that as humans we are nothing more than frequency suppliers and batteries for a mega quantum system that works with the human DNA!

The world that we have lived in until recently is all about clubs and groups that are superior and have all the rights. We are talking about a new world but also about another war of the mighty of the earth.
BUT........

Well yes, because everything is pointing to "Xirtam" being the mighty one of the mighty! The self-thinking and acting quantum computer which is linked to all systems around the world and already controls the world manipulation! Aren't we already under this powerful system? And is this why many powerful people withdraw into various worlds that can be made visible by that system? Is it not the case that they are draining our energy on a daily basis? Do we no longer need pleasures simply because we are energy sources?

The more you dive in and the deeper you go, the more you find yourself in what appears to us to be a dreamworld, but actually is the master control of our present life.

6-8
Dreamworld

I would like to make this statement

As a writer, I have described many things in my life, and many of them are presented as fantasies or lies. Although I have experienced, seen or experienced everything myself. Yes, I live in a dreamworld, according to others or in a world that does not exist! However, a lot has already come true, been proven or is in the process of being revealed! The dreamworld seems to be a reality and people have fallen silent.

All my life I have experienced that I see things that others do not. I sense them, I see them and I am in those worlds. The moment it became clear to me that we are not the only world on this globe and the data came in from the world below us, I understood many ancient writings and claims by people who have seen images but interpreted them completely wrong. The fact that distances, speeds and time all came together and that in energy, opened up the world of limitless possibilities.

All the energy is being wasted in order to travel to the moon or to live on Mars, even though this has been going on for an immeasurably long time now and we can see the evidence of it. And yet this information is being withheld and hidden to avoid giving us the impression that we are much further advanced than we are being made to believe.

The entire programme of travelling into space is to generate financial support for the real projects in progress behind the scenes. Those groups that are involved are far from anything earthly and are dealing with issues that are mega beyond of what we can imagine.

I have described all this already but I am quoting it now, because if there were no time restrictions, this information would be there for us too! "Time" is what makes us believe that we are still living in the technical Middle Ages! For that is what people have been doing for a long time. For that is what people have been doing for a long time. Much is denied and covered up for the outside world, and if you start talking about it, you are either dreaming or you are a fantasist who is just making up stories.

Since we, as ill-informed people, can observe but also record more and more things, stuff like large flying objects and activities on the sun, for example, are now coming out. These are no longer made-up stories, but recorded facts. In the same way, time is revealing more and more secrets and, due to newer techniques, is proving to require many calculations.

" People we are living in the world of quantum/DNA and
  Artificial Intelligence!"

6-9
Master control/ Life

Everything under control, is it the future or is it the present?

Since 2019, a lot of manpower has been used to get humanity on the same level. Wars faded and people were no longer set against each other by country or group, but instead we now have the individual acts of hatred and envy throughout the world. By now, if you have read carefully the previous parts of the book, we know that it is not about human intervention anymore, but that we are really being controlled and directed by simply put, "a computer". People laughed when the first films came out where robots were in charge, but we were confused when we saw the Matrix cycle, which in reality was the reference to the Xirtam quantum system/era.

Many wondered why Bill Gates had gone into healthcare worldwide, but people also suspected that it was about money or power. In the past, we have not only seen dark clubs being opened up, but also well-known high-ranking members of the system being pilloried, murdered or vanished. We also notice that without any resistance, governments surrender to a higher power. We can also see that the people are hardly protesting despite all the freedoms being taken away from them!

All this is placed under some pandemic and people act according to the rules that apply to the large groups and are ignored by a selected group! For a long time, there were many riddles of global 5G deployment, experimental vaccines that affect your RNA and eventually your DNA, or simply consume it! At the same time the large-scale repression where laws, treaties and rules no longer apply.

I mentioned 10 years ago that the law books had been rewritten. That there are other rulers was pretty clear and that IBM initially wanted to control the world via Verichip was also already established.

For a limited number of people, travelling between dimensions is not a problem, neither is the modelling of energy. Since technology is in control, the question is whether humans are still needed, apart from being drained through DNA and frequencies. "Xirtam' is the name of the system that controls the world and we know that China has the most powerful quantum computer and it can use any computer, quantum or not, to build up its know-how. Moreover, it is no coincidence that the whole scenario of the Corona scandal started in China. You have already read about this in a previous chapter.

Was this a journey to some future or was it the way a quantum computer works, simply travelling in the different dimensions according to the quantum principle?

Everything is pointing up the fact that we are not one of those groups that now run the world, but the absolute power is a computer! Then the question arises as to whether this was done deliberately or whether the quantum computer is becoming too autonomous. Considering their behaviour, the authorities, police, army and controlling groups are performing too many orders against humanity! Beating your fellow man to death and putting your other fellow man into dangerous actions is not human thinking any more, but rather the probability calculations of a computer called Xirtam.

The world seems to be ruled by a computer that knows so much about each person that if they do not listen or refuse orders, they will be pilloried! There is clearly a policy of extortion and frightening, or threatening those who have their eyes opened. Dealing with death is normal and an overdose is exactly what it looks like when manipulating DNA/RNA and switching off the frequencies needed for life. We are living under a regime of a computer that defines our world and controls our lives.

It was known for a long time that we are more and more under the spell of technology and it is being proven time and again that we cannot live without it! Like I mentioned before, we are not only flooded with films, images and articles, but we are also being bombarded with radiation, frequencies and manipulated food, produced by various Nano technologies!

Unaware of it, we have been fed by technology for 30 years now, and technology has taken hold of us all, part by part. The mightiest of them hardly ever use technology and they leave all the electronic work to others!

For example, the children of Steve Jobs did not have electronics around them. But also, over intellectuals hardly work with electronics! In other words, via electronics Xirtam comes into everyone's life and uses the information to expand its know-how!

Also remarkable is the rise of Mr Musk who apparently has an endless purse of money and who also manages to bring the world under one umbrella!

In the case of Bill Gates, we also see that he suspiciously stays away from various electronics and has been leaving everything to his "successors" in the electronic world for quite some time!

So what is it, that the 3 biggest ones in the world who are into electronics themselves but are staying far away from it?

You can see clearly how, in recent years, world leaders have surrendered all their power to the unknown, carrying out their tasks like sheep.

All the signs point to an "Autonomous" or "Self-determining" independent body which can be found in today's Xirtam!

In that case, is the pandemic not the controlling body and are the secret meetings not the discussions with Xirtam about which measures, sanctions, should be taken? Where is your true leader, who is apparently a computer and scatters his commands in every country accordingly to the reactions of the people?

Victims of vaccination will be the first to disappear. We can already see this with the old "useless" people, who according to quantum calculations die after a second injection! And wasn't this already described in timeless scriptures? Has Xirtam taken this wisdom as a guideline?

Therefore, is the absolute ruler not human, not a reptilian as is widely held, but rather a computer?

EVERYTHING POINTS TO THAT!

And as everything is directed by a computer, it is also clear that we can go into several dimensions once we have hacked the system! This happens on a regular basis and can be seen. Also, it is no longer important whether the entire UFO and history of mankind, is displayed in a manipulated way. That is why mankind can easily get to a lot of information already!

This happens on a regular basis and can be seen. Furthermore, it is no longer important whether the entire UFO and history of mankind is presented in a manipulated way. That is why mankind can easily have access to a lot of information already!

Or is this some programme that we are seeing?

This is where we come to a point of no time no distance.
A manufactured life?
All our lives we have been told that where there is a will there is a way.
Focus on something and it will reveal itself.
A déjà vue.
A purpose.
Every person creates their world.
Dimensions.
Believing.

Not only hearing voices, but also seeing images, is a proof that your frequencies are being manipulated. Through those frequencies your normal image, what you are allowed to see as a human being and what is held before you, has been disturbed! You are seeing or hearing other worlds/dimensions. For a moment your whole screen is disturbed and you get a glimpse of what is really around you. Just think of the scene with the red and blue pill in the Matrix, namely the world of computers or the real world! It was so-called science fiction but later they told us it was a documentary! And then this should be the film that tells us by whom we are really governed. It is becoming more and more clear that the world is not controlled by humans nor by reptillians, but for many decades we have been controlled by a computer which is so advanced that it thinks, acts and works on its own and does not need humans to continue! An autonomous machine that runs its own course!

Everything points to the fact that we live in a programme that we can manage a bit with the limited possibilities we are given. It all depends on your own sensors to what extent you will behave and live outside the normal programme. Now that all is controllable, it is clear that Xirtam has gathered its full "starting power". I call this 'starting power' because it is only the beginning, as at a certain time there will be no need for intermediate messengers and everything will be imposed through this one power.

Chapter 7

Final

7-1 Time gone, future gone?

Based on a human understanding, this would be a true statement, but the statement is based on the phenomenon of "time"! Since we know that time is a humanly imposed limitation, it therefore is not. Because time has gone, there is no future? Because time does not exist, the future is the now and can thus be created now without any problem. With everything on one line, it is simply a matter of going left or right from this line. We can say that you have to look directly into you and around you now, for that is where everything is happening. This can be done by opening up into the total energy, which then does not limit you and allows you to experience the many worlds but also to consciously accept them.

The future you want to see or live in has always been there, you just haven't opened up to it and therefore it can't be seen or experienced. Operating with very sophisticated equipment is about simulating that what you want to see. By doing so, one thinks one can then enter into many worlds, which have either been or are yet to come according to human concepts.

Being one with the total energy, you will clearly see everything happening around you. Once you have mastered this gift, you will see that there are some people who are further along in life than the masses. Through the evidence of the present time, we also see that many people have completely lost their way. Also those screaming beings who are so full of freedoms, while they already possess it, don't know where to look for it, let alone work with it!

Let's go through some questions/answers concerning the perhaps daily freedoms that are being held back:

Do you have to work?	No
Do you have to live like someone else?	No
Is there anything really stopping you from being free?	Yes

YES, because you are blocking yourself but also because you are full of excuses and 'but-thoughts'!

There are thousands of questions with the same answer.

If you don't work, you don't have money!
True, but do you need money?
Are you prepared for what is to come?
Do you trust totally that you will not disappear before your path is finished?
Yeah, some people die of lack of food but others live on and find a way.
Everyone is afraid of living short, whereas if there is no time there is also no short or long life!

An example
Circumstances forced me to quit my job at 47. There was no income from that day on! But in spite of that, opportunities presented themselves so that now at the age of 67, I still have a good life, not being hungry and having what I want around me! I have confidence that every day will give me whatever I need to go on.

When faith is there and you know that good will come to you, then those paths will be opened. These paths were already there for a long time, but I couldn't see them yet! There is an important fact in all this, and that is the word "have confidence" in the good.

7-2
Is there a true world?

Do I now live in a dreamworld?

No, absolutely not, and not every day is a day you would like to repeat. There are moments when there is no way out, let alone an answer. Then all of a sudden, simple things happen and new paths open up, which I then take. Whatever my common sense tells me, I go for it I know that if I take it with full conviction, there is something good at the end of it. In life, it is not about following your mind, but your feeling! We also see this in nature, and both the plant and animal world work by instinct. Hence, the most important thing in the world is to be guided by your feelings. The brain is there to sort out certain things that are theoretical. For a long time a computer could only work with that. It was the pure numbers that determined what to do and which decision to take. However, since the computer has taken the quantum side, a form of feeling can also be detected in a computer. Simply because, when analysing any human behaviour, it came across strange situations that cannot be explained through logic. The power of today's quantum computers therefore lies in thinking outside the box.

Thinking back for a moment to what I was experiencing myself, there were moments where I found myself that were inexplicable and the brain seemed to be in a state of confusion. For example, it wasn't long ago that I thus experienced, as I described earlier, absolute nothingness/blackness and I saw death. Then a computer starts to wonder what situation it got itself into. So I asked my guide what to do with that, and he told me it needed to be seen so I could get out of the very deep hole I had fallen into. Seeing death is a "must seen" fact in order to review what life is. It brought me a lot of pain and yes, you might get scared of that, but being scared is useless because what you are scared of will never turn out that way.

So:

What are you really afraid of?

Death?

Well, it is already within you! Life cannot exist without death!

Next, another strange question:

Do you know for sure that you are not already dead?

Who is to say if you are alive?

And how do you know you are alive if you don't know what life is?

Another series of questions that every brain has a hard time dealing with, because once again it is time to start thinking outside of all guidelines!

Perhaps life is not one energy field but multiple fields, which then appears to be a life? All is playing out in such a way to supply energy for the entire world of energy. That is something that emerges in every scenario, whether you ask a computer or a human brain mass. Because of the playing of that energy bubble, you and all of us believe that we are living one life.

And then there was a needle and it pricks that bubble and it's gone!

Is that the end of the whole story of your life?

So, what do you want me to respond to you on the question posed above, "Is there a true world?"

Because, let's be clear, what is the true world and how can you see that it is the true world? As a bubble of energy?

I personally do not see life as a game played with creatures of flesh and blood, for what is flesh and blood and what is true, what you see in nature and all the "life" that is around you. Isn't it like an energy movie, a hologram which travels through many dimensions and where a lot of energy is generated to make all that is impossible seem great! As I see it, we live in an environment of energy that we create and want to see, and a good example of this is love.

To a couple in love, the world is a true paradise but this same world others see as hell as they walk across the same square or even live together in the same house. It was a great party, the other looks and says "this is a lousy party". So, what makes me see it differently than the other person? All because we all live in our own world of energy, where everyone works and interacts with different energy and therefore interprets feeling, seeing, living and learning all differently! Your energy bubble works with different frequencies than the one of your neighbor, but you also approach things differently than another.

There is a great example in the world of energy. People would like to heal other people or help them from their misery. Your energy bubble cannot heal another's, because it will not allow it! There is no point in healing someone if you are not one with the other person and his energy. Moreover, that person you want to heal has to do it himself in his own energy bubble. One can never be in the same frequency vibes as another! Provided... Yes provided, you are completely one with that person and both energies are merging together for that purpose. That particular journey. Of course, healing is possible but after a while either the disease comes back or a worse disease rises! Merging of two different energies to heal is only possible for the highly gifted, not for those who have attended a course and have seen the light and have even let money play a role.

Healing others makes no sense as it is done in 99.8% of today's way. Molding in another's energy bundle requires an approval and cooperation from the world of energy of both parties. Not a simple yes, or some money from the patient but purely from the world of energy that determined this.

So, this is how we see "what is the true world"?
The true world is nothing more and nothing less than a bundle of energy which, per energy balloon, allows people to live their lives. Developing your energy, there are those who are much further than others. Then again,

in life, each energy balloon represents its own world with its own energy in which you can determine how far to go in that world of energy.

7-3
Mankind as an everlasting machine

In this part of the total explanation of mankind, we go one step further.

Is mankind an everlasting machine?

No and yes!

If one considers man of flesh and blood "No", it is then a machine that depends on his/her energenius task on this planet. That may be, an existence of a few minutes and others more than 100 years! But then again, there is no time so what will be the difference of that one minute or those 100 years? We will look at that later.

If one thinks of every living being as energy then the answer is "yes". After all, energy will not decay and so neither will the energy of any living being on earth. Native Americans see it as a reincarnation in any form. This reincarnation is based purely on being human/animal.

People still like to see a fixed image as another person, animal or picture than they start believing in pure energy. Consider all the images of gods and the mighty of the earth. People waste entire lives believing in a fictitious being, rather than beginning to see that everything is determined and guided by energy. The reason is that there are few who see energy and are able to unite with it. But there is no god either, at least no one has ever seen him! And yet, people follow those earthly divine rules from a Bible, Koran or whatever scripture they have. The gods are presented to people in the form of images, and sacrifices are even made to save and clean themselves! And yet, it's all the same as believing in ghosts. Ghosts one did see regularly! Now imagine that one starts believing in ghosts! Then again there will be priests who will just exorcise the demon without knowing what they are doing.

7-4
The world of energy of devils and spirits

Continuing for a moment about a demon or spirit. They are usually pictured as beings of evil in the energy. But in reality it is a disturbed energy mostly created by the church! The Church itself has created these beings and then portray them as demons or worse, as a representative of the devil. Do remember, every demon (energy balloon) has something against that cross and that cross has been distorting that world of energy. Mainly by previous abuses or even disturbing entire energies through dark ways. For the "devil" to then try to upgrade his/her energy is not at all abnormal.

The church itself has created these forms over the many thousands of years. And so, each one has created its own world. That there have always been stories and also true cases is normal because an energy will not allow itself to be destroyed! It is odd if you then think you can overcome a devil or evil spirit, while you cannot understand the even less affected energies.

People will have to realize what life is before they think they can do anything about it. One ferments, one experiments with frequencies, but also via Nano and undisclosed techniques. One is angling, one is messing around in the world of energy and this is causing the damaging of many energies and in some cases even in very serious forms.

Think for a moment of an atomic bomb, which is a destructive energy. But now I'm going to redirect some connections in that atomic energy and it has created a projectile that we humans can't control. That' s what we already have with those artificial intelligences where a quantum computer is managing everything! Because it is no longer humans who are in control! I neatly write "in control" but the truth is that mankind has never been in control and there has always been an energy that made and still makes the rules!

In the present corona era we see this clearly and I don't think I need to post any more examples from that period. People worked against every set of human rules, treaties and laws. In spite of that, they were implemented and man killed his fellow man! It happened because Xirtam put quantum thinking on it and stood far above humans and then came up with numbers that put the energy on self-destruction. Thankfully still in limited measures. People are calling the endless triggering of humans "an experiment," but no one is talking about what kind of experiment it really is, since Xirtam won't release it.

7-5
Is there any difference in that "1 minute of life" or "100 years of life?"

To humans yes, to energy no! We are done, aren't we?

Well, man thinks in time and yes then 1 minute of life is a bit shorter than 52,596,000 minutes!
Again, you see the ridiculous numbers coming up!

When we look at it in energy, it is nothing more than between the blinking of the eyes.

In the CERN in Switzerland, they are noticing that they will have to revise all numbers, since they are witnessing that there is more after the 'nothing' and also more after the absolute. With the release of Xirtam they will also see that they are working with dangerous toys, since they are entering a world of energy that they have never seen, let alone participated in.

Indeed, at the moment one is still nobly trying to find the limits (or at least so one pretends) but limits exist only in the world of numbers and not in the world of energy. That is why a human energy is unpredictable, but also the energy of a stone, leaf, bird or water.

Let's take a closer look at water: The wonderful thing about water is that one has absolutely no idea about the way in which this medium exists via the energy. As for water, it is like fire, light and shadow, one of the forces that one places in an ivory tower and is still far from the 'examined'. That now the factor 'nothing' has been added is a problem, because 'nothing' is not nothing, but something!

7-6
No past, no present

It is true that in the real world of energy there is no past and present. You are only, when you are engaged in energy, engaged in where you are in now. This is so because there is only one purpose and that is to generate energy for the whole. Energy is what keeps everything alive but energy also is that which puts all into work. It's not a matter of "we're just messing around". No, even that bum, that addict, that unemployed person or that person going to work every day are doing the same thing.

Being alive is generating energy and that's what we do, day in and day out.

So yes, everything points to the fact that we are simple battery chargers and some see a little more around that battery than others. It is getting more and more clear that we are in the service of one entity and that we are not just messing around in this world. Nature is clearer and more structured but also better organized. Man is the chaotic one and that creature who does not know what he is doing.

It's nice to see, that as humans we cling more and more to technology and we are seeking our salvation there! It is technology that has brought us to our present state! We are lost but also lost the source, we are wandering and we cling to the very thing that brought us to this state!

Looking through a different lens and allowing your eyes and your brain to register what is really going on, we realize that it doesn't matter what was then, what is now and what is yet to come. So yes, for human thought it is as if we are living from one minute to the next but in the world of energy there is only a NOW and that now is what you are creating! People are telling you many times in your life to "live now", and what do you do? You glance at your watch what the time is and you have lost that "now" moment again!

Time really is the greatest limitation man has imposed on himself. The calculations don't add up and man clings to a theory that only causes limitations.

7-7
Where shall we go from here?

Now that we know that there is no time and that time is a plaything of a system and that the system is nothing more than "bits and bytes", perhaps it is wise to start seeing things differently. Yes, you can make your life a hell but you can also make your life a paradise! The other thing we can do is to look for the true source and to find out if there are possibilities to contribute more to the whole. The fact that we remain batteries is abundantly clear, that is just life but we might look to find out more about which kind of a battery! It is obvious that we are talking about one whole whether it is here, there or elsewhere in the universe. That is how nature works but also how the rest of the whole works. People are here for each other and every little particle has a purpose, a function!

We also have the point that not everything we see, feel and know is based on energy! Is it perhaps an image that is held before us? So, does that explain why we have so many empty images now in front of us and what appear to be people are just holograms? Is this what people see when someone is being shifted? Is this what the as-labeled reptiles in front of us adapt or reveal a true image for a moment? Is this what you feel when you talk to an empty identity?

All blends together when one starts to see life in energy and everything seems to fit when one lets time and distance fade to those proportions it has always been, sun up, sun down! From a simple fact we have dug our own murder-pit and the striking thing is that the whole Artificial Intelligence and an Autonomous computer has made a huge point of that and has addressed it to impose mega restrictions on us.

7-8
And then!

And then it was silent....

Silent, because we are all shocked that we as humans are not as superior as we claim to be and have imagined. On the contrary, that grain of sand in front of us seems to have the same function as we humans do, only we have the additional goal of making life difficult for each other and living according to self-imposed limitations! Mankind has contributed little to the whole. All the signs are that humanity does not know how to work in a whole, not to mention work with it.

Because of the way we think, we believe that we are "great" and have made it far. Although we see that thousands of years ago there were high civilizations with much more wisdom and many more ways of living, people now think that a designed machine is going to make up for everything we actually lag behind. However, this machine that we call computer is a product of that main machine that has been determining a lot for a very long time and has been making humans too busy with themselves and with their ego.

The main quantum computer is an ancient and highly intelligent and self-directed machine. Fully working according to the world of energy and is completely Autonomous and Artificial Intelligence.

We should be quiet for good reason, because we are really a micro particle of the whole. Also, people don't realize that even though the whole uses our energy and our frequencies, we are not needed if we ruin our whole energy. As a result, very young people are already dropping out and others, of whom many think they are 'useless', are protected and go on living. It is not that it is measured by what humans think is right, one is dealing with a machine that has refined its algorithm so much that it knows exactly what it needs.

Conclusion

In this book we have discussed the phenomenon of 'time' quite extensively and finally we have arrived at the core of all life. By eliminating time, distance and human thinking many worlds have been opened up but also energies have been released which many have surely never experienced, felt or seen.

Please allow me to make it clear once again that strange things have happened to me during the writing process and I had to write two blocks all over again because my 'word program' spontaneously erased them completely after having saved them. At first, I was really angry because you spend time with a book, let information come into you, and the next day it's gone! Angry because the technology thought differently from what I had written. Later on, I realized that strange forces had occurred during the writing of this book, and other than the disappearance of text, it has never taken me this long to write a book and also made me see many things that were new to me.

This is a book full of surprises because it has a subject that many people certainly will not understand. I do hope, however, that one day people will have a different view, and that the new world, which is always seen in a negative light, will reveal itself to them.

The world of energy and where everything is possible in the "now", without limitations, without hate and envy and without a mean thought or competition. Common sense, called feeling, is about time that we start using that and then we have a chance as humans to create a beautiful whole together.

Be aware of the world around you, feel, look and touch it as it really is.

John

Additional Information

In The Gardian March 24, 2014, there was a scientific piece posted about proteins supposedly affecting the brain. Now 8 years later it is being put into practice in humans via the Covid vaccine and activated by the 5G frequencies.

So we see that certain strategies have been laid out for a long time and we, as humans are that battery that the Artificial Intelligence (Xirtam) needs.

The published article:

Genetically engineered 'Magneto' protein remotely controls brain and behaviour

"Badass" new method uses a magnetised protein to activate brain cells rapidly, reversibly, and non-invasively

Researchers in the United States have developed a new method for controlling the brain circuits associated with complex animal behaviours, using genetic engineering to create a magnetised protein that activates specific groups of nerve cells from a distance.

Understanding how the brain generates behaviour is one of the ultimate goals of neuroscience – and one of its most difficult questions. In recent years, researchers have developed a number of methods that enable them to remotely control specified groups of neurons and to probe the workings of neuronal circuits.

The most powerful of these is a method called optogenetics, which enables researchers to switch populations of related neurons on or off on a millisecond-by-millisecond timescale with pulses of laser light. Another recently developed method, called chemogenetics, uses engineered proteins that are activated by designer drugs and can be targeted to specific cell types.

Although powerful, both of these methods have drawbacks. Optogenetics is invasive, requiring insertion of optical fibres that deliver the light pulses into the brain and, furthermore, the extent to which the light penetrates the dense brain tissue is severely limited. Chemogenetic approaches overcome both of these limitations, but typically induce biochemical reactions that take several seconds to activate nerve cells.

The new technique, developed in Ali Güler's lab at the University of Virginia in Charlottesville, and described in an advance online publication in the journal Nature Neuroscience, is not only non-invasive, but can also activate neurons rapidly and reversibly.

Several earlier studies have shown that nerve cell proteins which are activated by heat and mechanical pressure can be genetically engineered so that they become sensitive to radio waves and magnetic fields, by attaching them to an iron-storing protein called ferritin, or to inorganic paramagnetic particles. These methods represent an important advance – they have, for example, already been used to regulate blood glucose levels in mice – but involve multiple components which have to be introduced separately.

The new technique builds on this earlier work, and is based on a protein called TRPV4, which is sensitive to both temperature and stretching forces. These stimuli open its central pore, allowing electrical current to flow through the cell membrane; this evokes nervous impulses that travel into the spinal cord and then up to the brain.

Güler and his colleagues reasoned that magnetic torque (or rotating) forces might activate TRPV4 by tugging open its central pore, and so they used genetic engineering to fuse the protein to the paramagnetic region of ferritin, together with short DNA sequences that signal cells to transport proteins to the nerve cell membrane and insert them into it.

When they introduced this genetic construct into human embryonic kidney cells growing in Petri dishes, the cells synthesized the 'Magneto' protein and inserted it into their membrane. Application of a magnetic field activated the engineered TRPV1 protein, as evidenced by transient increases in calcium ion concentration within the cells, which were detected with a fluorescence microscope.

Next, the researchers inserted the Magneto DNA sequence into the genome of a virus, together with the gene encoding green fluorescent protein, and regulatory DNA sequences that cause the construct to be expressed only in specified types of neurons. They then injected the virus into the brains of mice, targeting the entorhinal cortex, and dissected the animals' brains to identify the cells that emitted green fluorescence. Using microelectrodes, they then showed that applying a magnetic field to the brain slices activated Magneto so that the cells produce nervous impulses.

To determine whether Magneto can be used to manipulate neuronal activity in live animals, they injected Magneto into zebrafish larvae, targeting neurons in the trunk and tail that normally control an escape response. They then placed the zebrafish larvae into a specially-built magnetised aquarium, and found that exposure to a magnetic field induced coiling manouvres similar to those that occur during the escape response. (This experiment involved a total of nine zebrafish larvae, and subsequent analyses revealed that each larva contained about 5 neurons expressing Magneto.)

In one final experiment, the researchers injected Magneto into the striatum of freely behaving mice, a deep brain structure containing dopamine-producing neurons that are involved in reward and motivation, and then placed the animals into an apparatus split into magnetised a non-magnetised sections. Mice expressing Magneto spent far more time in the magnetised areas than mice that did not, because activation of the protein caused the striatal neurons expressing it to release dopamine, so that the mice found being in those areas rewarding. This shows that Magneto can remotely control the firing of neurons deep within the brain, and also control complex behaviours.

Neuroscientist Steve Ramirez of Harvard University, who uses optogenetics to manipulate memories in the brains of mice, says the study is "badass".

"Previous attempts [using magnets to control neuronal activity] needed multiple components for the system to work – injecting magnetic particles, injecting a virus that expresses a heat-sensitive channel, [or] head-fixing the animal so that a coil could induce changes in magnetism," he explains. "The problem with having a multi-component system is that there's so much room for each individual piece to break down."

"This system is a single, elegant virus that can be injected anywhere in the brain, which makes it technically easier and less likely for moving bells and whistles to break down," he adds, "and their behavioral equipment was cleverly designed to contain magnets where appropriate so that the animals could be freely moving around."

'Magnetogenetics' is therefore an important addition to neuroscientists' tool box, which will undoubtedly be developed further, and provide researchers with new ways of studying brain development and function.

Reference
Wheeler, M. A., et al. (2016). Genetically targeted magnetic control of the nervous system. Nat. Neurosci., DOI: 10.1038/nn.4265

Video
https://youtu.be/iHTpJNSNFlc

Credits

Picture data

Central	The tokamak design of the fusion research facility in Culham. Science / business
Clock	"Klokkenwinkel", Lier
Conversation	Triplets Have Conversation With Unseen Figure,
Quantum.	Quantumcomputer Dubai
Technology	Facebook

If there are other credits of the photos, please let me know, because unfortunately I did not find them.

But from my heart and soul, thank you for your illustration/image.

John

John Baselmans wrote several books.
These books can be ordered on the website;
http://www.johnbaselmans.com/Books/Books.htm
The published books are:

John Baselmans Drawing Course	ISBN 978-0-557-01154-4
The secrets behind my drawings	ISBN 978-0-557-01156-8
The world of human drawings	ISBN 978-0-557-02754-5
Drawing humans in black and white	ISBN 978-1-4092-5186-6
Leren tekenen met gevoel	ISBN 978-1-4092-7859-7
John Baselmans' Lifework part 1	ISBN 978-1-4092-8941-8
John Baselmans' Lifework part 2	ISBN 978-1-4092-8959-3
John Baselmans' Lifework part 3	ISBN 978-1-4092-8974-6
John Baselmans' Lifework part 4	ISBN 978-1-4092-8937-1
John Baselmans' Lifework de luxe part 1	
John Baselmans' Lifework de luxe part 2	
John Baselmans' Lifework de luxe Curriculum	
Eiland-je bewoner Deel 1	ISBN 978-1-4092-1856-2
Eiland-je bewoner Deel 2	ISBN 978-0-557-00613-7
Eilandje bewoner - Luxe edition	ISBN 978-1-4092-2102-9
Eiland-je bewoner Bundel	ISBN 978-0-557-01281-7
Mañan	ISBN 978-1-4092-8949-4
He oudje leef je nog?	ISBN 978-1-4092-8482-6
De wijsheden van onze oudjes	ISBN 978-1-4092-9516-7
Makamba	ISBN 978-1-4461-3036-0
Onze Cultuur	ISBN 978-1-4475-2701-5
Ingezonden	ISBN 978-1-4092-1936-1
Moderne slavernij in het systeem	ISBN 978-1-4092-5909-1
Help, de Antillen verzuipen	ISBN 978-1-4092-7972-3
Geboren voor één cent	ISBN 978-1-4452-6787-6
Pech gehad	ISBN 978-1-4457-6170-1
Zwartboek	ISBN 978-1-4461-8058-7
Mi bida no bal niun sèn	ISBN 978-1-4467-2954-0
Curacao Maffia Eiland	ISBN 978-1-4478-9049-2
De missende link	ISBN 978-1-4710-9498-9
Curatele	ISBN 978-1-4717-9319-6
Curacao achter gesloten deuren	ISBN 978-1-304-58901-9

De MATRIX van het systeem deel 1	ISBN 978-1-291-88840-9
De MATRIX van het systeem deel 2	ISBN 978-1-291-88841-6
The hidden world part 1	ISBN 978-1-326-03644-7
The hidden world part 2	ISBN 978-1-326-03645-4
Geloof en het geloven	ISBN 978-1-326-28453-4
Dieptepunt	ISBN 978-1-326-71278-5
Namen / Names	ISBN 978-1-326-81898-2
Drugs	ISBN 978-1-326-84325-0
De protocollen van Sion 21ste eeuw	ISBN 978-0-244-61655-7
Verboden publicaties	ISBN 978-0-244-91960-3
De maatschappelijke beerput	ISBN 978-0-244-36559-2
Achter de sociale media schermen	ISBN 978-0-244-14015-1
Project Corona/ COVID-19	ISBN 978-1-71664-848-9

Omnis 1	ISBN 978-0-244-10848-9
Omnis 2	ISBN 978-0-244-40848-0
Omnis 3	ISBN 978-0-244-70848-1
Omnis 4	ISBN 978-0-244-10849-6
Omnis 5	ISBN 978-0-244-40849-7
Omnis 6	ISBN 978-0-244-81855-5
Omnis Photos	ISBN 978-0-244-10859-5
Omnis Photos 2	ISBN 978-0-244-44015-2
Omnis Photos 3	ISBN 978-0-244-21863-8
Omnis Photos 4	ISBN 978-1-71664-569-3
Omnis Photos 5	ISBN 978-1-71664-567-9

Words of wisdom (part 1)	ISBN 978-1-4452-6789-0
Words of wisdom (part 2)	ISBN 978-1-4452-6791-3
Words of wisdom (part 3)	ISBN 978-1-4461-3035-3
Words of wisdom (part 4)	ISBN 978-1-4710-8130-9

The world of positive energy   ISBN 978-0-557-02542-8

Het energieniale leven   ISBN 978-1-4457-2953-4
Dood is dood "energieniale leven"   ISBN 978-1-4476-7213-5
Zelfgenezing "energieniale leven"   ISBN 978-1-4709-3332-6
Levenscirkel "energieniale leven"   ISBN:978-1-300-76189-1
Utopia "energieniale leven"   ISBN 978-1-329-51188-0
Vrijheid en liefde "energieniale leven"   ISBN 978-1-329-79390-3
Dimensies "energieniale leven"   ISBN 978-1-365-87087-3
Hologram "energieniale leven"   ISBN 978-1-387-72155-9
Het lang verborgen geheim "energieniale leven"   ISBN 978-0-359-70533-7
Quantiversum "energieniale leven"   ISBN 978-1-71657-634-8

NU   deel 1   ISBN 978-1-4092-7691-3
NU   deel 2   ISBN 978-1-4092-7736-1
NU   deel 3   ISBN 978-1-4092-7747-7
NU   deel 4   ISBN 978-1-4092-7787-3
NU   deel 5   ISBN 978-1-4092-7720-0
NU   deel 6   ISBN 978-1-4092-7742-2
NU   deel 7   ISBN 978-1-4092-7775-0
NU   deel 8   ISBN 978-1-4092-7738-5
NU   deel 9   ISBN 978-1-4092-7768-2
NU   deel 10   ISBN 978-1-4092-7708-8
NU   deel 11   ISBN 978-1-4092-7759-0
NU   deel 12   ISBN 978-1-4092-7661-6

Het dagboek van een eilandsgek   ISBN 978-0-359-85040-2